William M. Banks

Clinical Notes Upon Two Years' Surgical Work

in the Liverpool Royal Infirmary

William M. Banks

Clinical Notes Upon Two Years' Surgical Work
in the Liverpool Royal Infirmary

ISBN/EAN: 9783337162054

Printed in Europe, USA, Canada, Australia, Japan

Cover: Foto ©berggeist007 / pixelio.de

More available books at **www.hansebooks.com**

V. MITCHELL BANKS, F.R.C.S

to the Liverpool Royal Infirmary, Pr

TABLE OF MAJOR AMPUTATIONS.

No.	Sex.	Age.	Amputation.	Disease or Injury.	Complications.	Result.	Cause of Death.	Date.
1	Male, ...	56	Shoulder-joint.	Enchondroma of humerus		Recovered.		
2	Male, ...	48	Do.	Railway smash.	Serious injuries to chest.	Died.	Chest injuries.	60 hours.
3	Male, ...	34	Upper third of arm.	Do.	Two compound fractures of skull.	Recovered.		
4	Male, ...	14	Upper third of arm and middle third forearm.	Machinery accident.		Recovered.		
5	Female,	9	Upper third forearm.	Traumatic gangrene.		Recovered.		
6	Male, ...	37	Middle third of thigh.	Railway smash.	Loss of blood during transport.	Died.	Shock.	12 hours.
7	Male, ...	23	Lower third do.	Knee-joint disease.	Spinal curvature and lumbar abscess.	Died.	Exhaustion.	9 days.
8	Female,	12	Do.	Do.		Recovered.		
9	Male, ...	32	Do.	Do.		Recovered.		
10	Male, ...	10	Lower third thigh and lower third of leg.	Railway smash.	Serious loss of blood and prolonged exposure to cold. Surgical scarlatina.	Recovered.		
11	Male, ...	49	Upper third of leg.	Tramcar smash.	Fracture. Extensive bruising of the opposite leg.	Died.	Delirium tremens. Sloughing of the flaps. Diffuse cellulitis in opposite leg.	10 days.
12	Male, ...	33	Do.	Railway smash.	Compound fracture of jaw and crushing of opposite foot.	Recovered.		
13	Male, ...	45	Syme's amputation.	Do.		Recovered.		
14	Male, ...	60	Do.	Ankle-joint disease.		Recovered.		
15	Male, ...	23	Do.	Do.		Recovered.		
16	Male, ...	44	Do.	Do.		Recovered.		
17	Female,	66	Mackenzie's amputation.	Enchondroma of calcaneum.		Recovered.		
18	Male, ...	50	Chopart's operation.	Intractable ulcer of sole.		Recovered.		

Amputation of the great toe, 3 cases; Amputation of one or more fingers or toes, 6 cases.

AMPUTATIONS AND EXCISIONS.

Four Fatal Cases of Amputation—From the accompanying table it will be seen that, out of eighteen major amputations, four cases proved fatal. Before taking note of any features of interest in the successful cases, it is well to observe what were the causes of the failures.

Fatal Amputation of the Thigh for Knee-Joint Disease.—Only one of the four fatal amputations was done for disease. The lad was at once the victim of suppurative destruction of the knee-joint and of antero-posterior spinal curvature. When he was ad-mitted, his left knee-joint was a bag of pus, his backbone was literally doubled up on itself, and in his left lumbar region was a huge chronic abscess connected with the spinal mischief. The knee-joint and the spinal abscess were repeatedly aspirated, only to fill again. Then they were laid open antiseptically. After eight months of treatment the spinal abscess had diminished to a mere sinus in the loin, from which exuded a little thin discharge; but the knee-joint had gone all to wreck, and gave the lad the most excruciating pain. At the patient's urgent request, the limb was removed just above the knee, to relieve him from his agony, but with small hope of his pulling through. Even if he had, the spinal mischief would still have been before us. He struggled on after the amputation for about ten days, and then died from sheer exhaustion.

Three Fatal Primary Amputations.—The other three cases were primary amputations for severe injuries. In one instance the patient died in about 60 hours after amputation at the shoulder-joint, from internal injuries produced by crushing of the left chest. In another, the man was run over on the railway about twelve miles from Liverpool, and lost a great deal of blood before any assistance could be procured. Then he was jolted in trains and cabs to the Infirmary, and arrived, in about three hours after the accident, so nearly dead, that it was only by hypodermic injections of ether and careful fostering of his strength, that he was tided over the night. By next afternoon he was so far recovered that amputation was performed through the middle third of the right thigh, the limb being irretrievably smashed. His strength was, however, too far gone to enable him to take advantage of the operation. The third case was that of a drunken cobbler, who was run over by a tramcar. He was a very bad subject indeed, with tissues completely saturated with alcohol. So that, although his left leg was cut off high above the injury, the flaps sloughed at once, he became violently delirious, and his opposite leg—the subject of a simple fracture, with severe general contusion—was seized with diffuse cellulitis, requiring extensive incisions. He had no stamina left to bear all this, and died exhausted on the tenth day.

In these four cases there was nothing to regret. Amputation was a necessity forced upon the operator in order that the patients might have the benefit of the few remaining chances left to them, and death resulted from causes over which surgery at present has no control, seeing that human beings are only endowed with a limited amount of *vis vitæ*.

Of the **Fourteen Successful Amputations**, eight were performed for disease and six for injury. Of the cases of disease two were done for enchondroma.

Amputation at the Shoulder-Joint for Enchondroma of the Humerus.—Samuel D. was a spare, placid man of 56, a chapel keeper. So far back as the summer of 1865 he was seized with a violent pain near the right shoulder, and after that there came a hardness and swelling at the top of the humerus, which very slowly increased. Very slowly, but very steadily, an enchondromatous tumour developed itself just beneath the deltoid. As it gave him no great inconvenience, he did not heed it much for many years; but by 1878 it had grown to be as big as a cocoa nut, so that, on attempting to raise the arm, it became locked against the acromion and limited movement, while pain of a severe character set in. In June of the same year (1878) the tumour was removed by cutting down upon it, and dissecting off the tissues from over it. As it grew from the outer surface of the upper third of the humerus, this was effected without difficulty. Then with a mallet and chisel it was cut clean away from the bone, and the surface from which it sprung was thoroughly scraped—a pretty broad surface by the way. I left no cartilaginous remains that could be seen. The patient rapidly recovered; but in the track of the wound a sinus or two persistently remained, leading down to the bone. After the lapse of about two years it became clear that the tumour was returning; and by the summer of 1881—three years after the first operation—it had attained an immense size, having taking a fit of growing during the last few months. It clearly arose from the same site as before; but now it filled up the axilla, and had even got beneath the great pectoral muscle. Pain and rapidity of growth demanded its speedy removal. But removal of a whole right arm at the shoulder-joint seemed such a dreadful thing that one was anxious to save a hand and forearm by carrying away, if possible, the tumour and upper part of the humerus, even although the upper arm might remain useless. The patient being made well aware that, in case of the failure of this project, there was nothing left but amputation, I attempted it. The incisions necessary to lay bare the tumour were very extensive, the chief one reaching from above the acromion half way down the outer side of the upper arm. With much trouble, and after the loss of a great deal of blood, the outer and upper surfaces of the growth were exposed, and the humerus was disarticulated from the scapula. Then, sawing through the humerus about an inch below the deltoid insertion, I attempted to dissect away the tumour from the brachial vessels and nerves. Here, however, most serious difficulty was encountered from their intimate incorporation with the growth; and at last, after a prolonged attempt, I was reminded by my colleague, Mr. Harrison, that the patient had plainly endured as much as he could, and that to make further effort might only lead to collapse on the table. I was reluctantly compelled to admit this, and so rapidly swept the limb away at the shoulder. So profound was the shock that a short time after the operation the temperature fell to 95°, and remained so for many hours. The operation was conducted antiseptically; and the patient, in spite of the loss of blood, made such a rapid recovery that, on the twenty-third day he left the Infirmary quite well, and remains so now, two years after the amputation. If the great vessels and nerves have been saved, although with the loss of the upper half of the humerus. But even a forearm is better than no arm at all. The case also shows that chiselling off cartilaginous tumours is not by any means a certain removal. The surface that was left upon the humerus, after the first removal of the tumour, looked perfectly healthy to the naked eye; but there must have been cartilage cells deep down in the tissue of the bone.

Central Enchondroma of the Calcaneum—M'Kenzie's Operation.— The patient was a healthy looking woman of 66 who, between three and four years before her admission, noticed a small hard lump on the outer side of the right calcaneum. This very slowly increased till a few months before she came to the Infirmary, when it began to make great strides. It became hot and red, and the skin over it gave way, giving vent to a pale yellow fluid. The tumour was globular in shape, and the measurement round the tip of the heel and front of the ankle was fourteen inches. On the outer side of the heel and ankle the skin was destroyed, and there was an ugly looking hole out of which a sanious stuff exuded. When the finger was thrust into this hole bits of loose bone could be felt, and, on one which was extracted, a layer of cartilage was found. The tumour was diagnosed as a central enchondroma of the calcaneum. As it was impossible to remove the foot by Syme's operation, owing to the destruction of soft parts on the outer side, M'Kenzie's method was adopted; and from the inner ankle, and such of the hard part of the heel as was available, a flap was obtained, which was turned up under the sawed surfaces of the tibia and fibula. It makes a very excellent stump, as the drawing

from a photograph—taken some months after the operation— shows. The enchondroma was found to have originated in the centre of the calcaneum, and to have expanded and then split up the bone in the course of making its way to the outer side.

Chopart's Amputation, upside down.—Peter Bourke, a baker, aged 50, thirty-four years ago had a terrible typhus, and, while in a workhouse hospital, had certain hot bottles applied to his feet, which took the skin of the greater part of the sole of the left one— a warning to nurses. This left a large tender cicatrix, which, as a result of a crush, broke out some years ago, and never could be properly healed. Under prolonged treatment in the Infirmary, it at last skinned over; but the man was barely at work again when it gave way, and he was as bad as ever, so that he came and requested that the foot should be removed. As it was desirable to save as much foot as possible, a sort of Chopart upside down was performed. The flap was made by dissecting up all the soft tissues on the dorsum of the foot, down to the roots of the toes. The sole was next cut straight across just behind the sore, and then as much foot was cut away as was necessary to allow the dorsal flap to come down over the face of the stump. This left the calcaneum, astragalus, and scaphoid. A very capital foot has resulted. One considerable objection to partial amputation across the tarsus has been the drawing up of the heel— which is certainly prone to occur—and which produces a tendency for the patient to walk on the front of the stump. This is in part to be accounted for by the loss of the fore-part of the foot, by reason of which the astragalus, deprived of its support in front, tends to slide downwards and forwards over the calcaneum. But, doubtless, it is partly due to the tendo Achillis dragging the heel up during the process of cure, which it readily does, owing to the extensor tendons being all cut. With the view of preventing this occurrence various devices have been employed, such as stitching these tendons to the sole flap; but the division of the tendo Achillis at the time of operation (which was done here) has, in my experience, always been sufficient by itself. By the time this tendon has united its opponents in front have fixed themselves permanently in the flap. Some three years ago I performed Chopart's operation upon a young gentleman for a gunshot wound of the foot. The tendo Achillis was divided, and the forepart of the foot kept well up during the healing process. So good a stump resulted that he is now an excellent cricket player, and recently won a prize at a long swimming match.

Two Amputations above the Knee-Joint for Disease were performed successfully—one on a girl of 12 and the other on a man of 37. Both were for disease of the articulation. As to whose amputation was performed, I really don't know. Such a number of gentlemen have lately been desirous of associating their names with amputations that, at any given part of a limb, you have Mr. Jones' amputation just *above* the superior tubercular process, Mr. Brown's ditto just *below* the said process, and Mr. Smith's and Mr. Robinson's modifications of the same *through* the process. For my own part I only know, and I only teach two things about amputations, and these are to make one flap longer than the other, and to saw the bone as low down as possible. In every amputation for disease of the knee-joint, I follow one plan, viz., to dissect up a very long anterior flap, the point of which may, if necessary, reach two or three inches below the lower edge of the patella, and to cut a short posterior. I saw the femur across just clear of the cartilage, and look at the section. If the bone is good, that will do. If it is not, I saw off another slice, and, if need be, another, until it *is* found to be good. It is difficult to see any practical use in keeping the patella. Sawing off its cartilaginous surface, and then trying to make it stick on to the cut end of the femur, may afford an operator of a mechanical turn of mind some extra amusement, but nothing more. It isn't the patella that is wanted, but the hard skin over it; and whether that skin covers the patella or covers the rounded end of the femur, is a matter of indifference, seeing that in no well constructed artificial leg is the whole weight of the body made to rest on the face of the stump. In a well made knee stump the long front flap should join the other a considerable way behind the end of the bone, and, speaking generally, the uglier and clumsier the stump at the time of the operation the better will it be six months afterwards. A neat, tidy, tight stump on the operating table generally results, after a few weeks, in the surgeon contemplating the cut end of the bone, and consoling himself by studying the process of its occlusion as witnessed by the naked eye. Concerning the soft parts, it is a matter of indifference whether the skin is riddled with holes or not, or how jelly-like the subjacent tissues are. Let the skin holes be cleaned up with a spoon, and let the jelly be cut and scraped away, and let the sinuses, which always run up on the inner side of and behind the femur, be purged of their flabby granulations, and let the whole thing be swabbed over with chloride of zinc solution. After that the most pus-soaked, gelatinous stuff will make as good flaps as the best. I have made capital stumps with material so brittle and rotten that, while making them, I have hardly ventured to turn the flaps very much back, lest they should break off. The way they soften down, so soon as the diseased joint is removed, is marvellous.

I suppose in this country it is really to Teale that we owe the principle of the one long flap to cover the end of the bone, while Spence, at the knee-joint, doing away with the trigonometrical survey prescribed by Teale, showed that the surgeon need only use his eye to measure his flap. But they knew something about it two hundred years ago. There is an admirable little treatise on amputation below the knee by Verduin of Amsterdam, published in 1696, in which he figures a long posterior flap and the resulting stump. The accompanying drawings are copies of his plates. The larger one shows a rather ingenious splint, intended to hold the long posterior flap in position.

Syme's Amputation.—Three cases were performed for disease and one for accident. The more one does this operation the more satisfactory is it found. As for Pirogoff's modification, I have only performed it in two or three cases, simply because I never could see that it possessed any notable advantages, while it necessitates our obtaining union between two surfaces of bone which are not always in a sound state. For cases of accident, where the bones are healthy, it does very well. But some people have a notion that, because the skin has been kept over the piece of bone that properly lies beneath it, it will bear more weight. It does nothing of the sort, for the very good reason which has been already urged, that it is not the patella or the calcaneum at all that are wanted, but the hard skin over them; and it does not matter over what bone this is planted. To show the value of even a fragment of the hard skin over the heel, one may allude to the case that was done for injury. The patient was run over on the railway, and his foot very much mangled. It was most important to make a Syme's stump for him, if possible; but the tissues were so much bruised that it was questionable if they would live. Nevertheless, the operation was done on speculation. More than half of the heel flaps sloughed, leaving only about an inch, or a little more, of the very back part of the heel skin. The parts were allowed to granulate and close up as far as they would; and then, after some weeks, the cicatrix was opened, the soft parts were stripped up, about an inch of tibia and fibula were removed, and this small bit of heel skin brought on to the face of them. The patient was seen the other day, able to work and equal to a five mile walk—bearing the whole weight of his body on this fragment of skin.

Syme was very particular about the mode of doing his operation, and, like all inventors, asserted that if his principles were not carried out to the last iota, all would go wrong. Nevertheless, every one knows the difficulty of dissecting back the heel flap, and the almost impossibility of doing so if the plantar incision is at all in front of the tuberosities of the calcaneum. But these tuberosities are variable in size, and sometimes only reach a little way forward, and in such a case the heel flap must be very short. During the past few years I have gone quite on the opposite tack, and have made the plantar incision well forward in the sole of the foot, so as to secure a large flap. This having been separated as far back as can be comfortably done, the tissues are loosened from the bone beneath each malleolus. Then opening the joint, and dividing the lateral ligaments, the back and lower surface of the calcaneum are steadily cut upon until the bone is cut clean out. In other words, in place of dissecting the fleshy cup off the bone from before backwards, the bone is almost entirely dissected out of its fleshy cup from behind forwards. The advantages of this proceeding are that the somewhat troublesome process of dissecting back the heel flap is simplified, and a large fleshy stump obtained, which the tendons very generally can move a little. This plan, although but little noticed in books, is the one usually practised by many surgeons in the South of England; but in the North and in Scotland, the original directions of the master are rigidly adhered to. The plan of making a small incision in the bottom of the cup, and inserting a drainage tube, is more than ever useful in the present days of antiseptics, because with an exit for blood or discharges thus permitted, we can endeavour to procure primary union along the whole line of incision without any fear of retention.

Boneing the Great Toe.—I have ventured to apply this name to a method of removing the great toe for disease of the metatarso-phalangeal joint. Three cases were done, and I have done two or three since 1881, the results of which have been in all respects satisfactory. A few years ago I diligently practised excision of the metatarso-phalangeal joint, but have abandoned the proceeding, finding that the time required for healing, and the trouble resulting from subsequent necrosis of the ends of the bones, were not counterbalanced by corresponding advantages. Nevertheless, it is quite worth while saving all the soft parts; and in order to do so, I have taken to the following plan. A single incision is made along the dorsum of the toe from about an inch above the metatarso-phalangeal joint down to the tip, thus dividing the nail in halves. Then the nail is wrenched off, and the soft tissues are dissected away on each side from the phalanges and end of the metatarsal bone, which bone is finally sawn through just behind its head. I have more than once lifted the bones and joints out entire, and have then split them longways with an old knife and a mallet, in order to show the students the difference between the diseased metatarso-phalangeal joint and the sound phalangeal one. It is quite true that the empty soft parts shrink back and shrivel up very much, but still a kind of great toe is left, which, although somewhat wobbly, fills up a vacuum in the boot, and gives greater firmness in walking. Besides, it does not look nearly so bad a mutilation. I think it is an improvement on the old clean-sweep way of amputating the toe, and only takes a few more minutes to do. The drawing is from a photograph of one of the cases taken about eighteen months after the operation.

Amputations for Injury.—Among the amputations for injury were some very remarkable recoveries. For instance, Robert Evans, a platelayer, aged 34, was knocked down at Edgehill Tunnel, and his right arm smashed above the elbow-joint, requiring removal three inches below the shoulder. In addition to this he had five scalp wounds. At the bottom of one was found a long fracture of the right parietal bone, with one side slightly depressed; and at the bottom of another was an almost identical fracture of the occipital. Antiseptics were employed for the amputation, and the head was shaved, and all the wounds treated antiseptically also. He had a rise of temperature of about two degrees during the first four days, and that was the only symptom he showed of anything being the matter with him. On the twenty-third day he was up and about.

Two double amputations were performed. In one case a boy of 14 had his arms dragged into some machinery in a rope factory, and my colleague, Mr. Parker, in my absence, found it necessary to remove one just below the shoulder, and the other in the middle third of the forearm. An excellent recovery followed. The other case was almost affecting in some of its details. Albert Smith, a young imp of 10, having found half-a-crown, ran away from his home at Miles Platting, near Manchester, in order to get to sea from Liverpool. He took a third-class ticket, and, being alone in the compartment, was leaning out of the window as the train was coming down Lime Street Tunnel. The door flew open, he fell out, and the train went over both his legs. This was in the evening. He had strength enough to crawl to the wall of the tunnel, against which he sat up and tugged vainly at the signal wires, which were within his reach, till cold and loss of blood rendered him unconscious. At 9.30 next morning he was discovered by the men who search the tunnel apparently dead. No wonder!—the night was the 14th January, 1880—one of the most bitter of an intensely cold winter—and all through the freezing night he had lain in the tunnel with his legs smashed to pieces. On his arrival in a blanket at the Infirmary, at 10 a.m., he was at first deemed dead; but, some flickering signs of vitality appearing, he was plied with restoratives, and, in the course of an hour and a half, some pulse could be made out at the wrist—notably after two subcutaneous injections of 30 minims of ether. By 10 o'clock he could utter a few words. He was placed on a mattress opposite a large fire, and literally cooked into life again. By 9.30 at night fair reaction had set in, and accordingly he was carried into the theatre, and got the smallest whiff of ether, while, with all the rapidity possible, one limb was amputated at the lower third of the leg and the other at the lower third of the thigh. The limbs were removed, rough dressings of lint soaked in carbolised oil were applied, and the patient was again in the ward on his mattress before the fire in less than fifteen minutes from the time of his removal. His recovery was undoubtedly greatly due to the rapidity with which the removal of the smashed limbs was accomplished, by which any prolonged shock was avoided. As if he had not already had enough to kill him, he had an attack of that peculiar fever which sometimes assails children after operations, and which bears such a close relation to scarlet fever as in most points to be indistinguishable from it. The rash appeared on the 20th day after operation, and was brightest over the backs of the hands and fore-arms. For some days his temperature ranged between 101° and 105°. Free desquamation ensued, the eyelids were œdematous, and for nearly three weeks the urine contained albumen—averaging one-tenth for part of the time. The flaps, meantime, opened up and refused to heal, and the bones protruded. After innumerable vicissitudes, the patient at last struggled through; and, some three months after the first operation, a slice of bone was removed from each stump, and they were made quite presentable. The longer leg was fitted with a Thomas splint, and with that and a crutch, he gets about with the greatest ease, and is in no worse plight than many a patient who has ònly lost one leg. He is now learning to be a clerk in a lawyer's office. The drawing shows the lad some time after the operation, and illustrates the manner in which the Thomas's knee splint is employed as an artificial leg.

There can be little doubt that the patient's recovery in this instance was largely due to the rapidity with which the removal of the limbs was effected. In the pre-anæsthetic days rapidity of operation was everything, and the man who could cut off a leg in the fewest number of seconds was the prince of operators. Not only has all this departed, but it is a question whether we are not running to the opposite extreme. Students see such operations as those for cleft palate and vesico-vaginal fistula, for which the patient is kept anæsthetized for an hour or two, and are apt to think that all patients may be kept *ad infinitum* on the table after the same fashion. I cannot help a certain fear that surgery, in some departments, is becoming too mechanical—that too much attention is paid to the planing, screw-driving, and sawing, without any regard to the kind of wood that is being used. The *vis vitæ* of the patient is not sufficiently regarded; his real and his surgical age, his occupation, his habits of life, the state of his internal organs, do not seem to attract that amount of attention before a capital operation which they once did. But, above all, the "rising" surgeon has difficulty in appreciating with proper keenness the condition of shock resulting from serious accidents. Formerly the screams of the agonized patient compelled the surgeon to hurry: now, having only a log to deal with, he is apt to proceed calmly— too calmly. He forgets that, although the patient is silent, and is not suffering pain, he is suffering shock—that every minute of anæsthesia, every fresh incision, every lost teaspoonful of blood is steadily diminishing his chances of survival. In a thigh amputation for smash, the fact of the patient being on the table twenty minutes in one case, or three-quarters of an hour in another, makes all the difference between his crossing the bar and sticking on it.

Thomas Beckett, a signalman, aged 33, was knocked down by a passing train, which went over his left leg and right foot, and threw him down among the levers, which cut him badly about the face and fractured his lower jaw. This occurred at Garston, six miles from Liverpool. His left leg had to be amputated below the knee, and the two outer toes, with their metatarsal bones, removed from

the opposite foot. The left leg was treated antiseptically. The right foot, being severely crushed, was not, but was merely wrapped in carbolic oil dressings. By the seventeenth day the leg stump was absolutely healed without one drop of pus. *Per contra*, the right foot began to slough, and a condition closely resembling the old-fashioned hospital gangrene set in, which nearly cost the patient his life, and certainly ultimately entailed the loss of a considerable portion of foot over and above what was removed at first. The patient's limbs were under the same blanket. Had the putridity of the right foot spread to the major amputation, nothing could have saved him. Nevertheless, with a condition in the foot requiring it to be buried in charcoal, and rubbed over with fuming nitric acid, the leg amputation close beside it—under its antiseptic cover— healed straightway without a trace of suppuration. If, in the face of what may be termed "natural experiments" such as these, some people still sneer at Listerism as a fad, further argument with them is futile. As regards the lower jaw, it had been fractured so that a wedge-shaped piece—carrying three incisor teeth— was practically lying loose in the mouth, and was lifted away. Now the curious thing was that this serious injury was effected without even a scratch visible on either lip. If one had been asked in a witness-box whether it was possible so to break a man's jaw as to inflict the injury just mentioned—without even leaving a mark on the soft parts around the mouth—it would have been hardly possible to admit that it could occur. So considerable was the lost piece that the fragments could not be well kept together— more particularly as we were unwilling to harass the patient with apparatus about his mouth while he was so ill otherwise. After the lapse of a couple of months, the maxilla was drilled on each side of the fracture, one of the drilling machines employed by dentists—which are driven by the foot—being used. A strong silver wire was passed through the holes, and, being twisted, drew the fragments together. Then the drill itself was pushed from the outside of the cheek through one ramus about the level of the first bicuspid tooth, then through the other and out at the opposite side. It was left in position, its ends being cut off and sealing wax put over them. Thus the fragments were kept in absolute apposition, and the patient could move them as one piece. The wire was kept in for two months, and the drill for six weeks, at the end of which time it began to work loose in the bones. After all it was found that too much maxilla had been lost to allow of bony union between the fragments, and all we could get was a firm fibrous bond. This serves the man's purpose very well for anything short of cracking nuts and sea biscuit. For drilling bones the dentist's machine is unrivalled, making the drill point go as if through soft wood. With hard bones, like the femur and maxilla, this makes all the difference between a neat job and a clumsy one.

The four cases just described were obviously of the gravest possible character. Indeed, the cases received at the Infirmary which are classed in the category of serious accidents are usually *very* serious. A large proportion of them are railway injuries occurring in the neighbourhood of Lime Street tunnel, or brought from country stations six to ten miles out of town. The latter are simply dreadful—the patients arriving in the most exhausted condition from loss of blood—from jolting in trains, carts, and litters —and from the agony of tourniquets tightly bound to restrain hæmorrhage. Were it possible to treat these cases on the spot, many patients would recover who are literally killed in the process of getting them into town. But what can the country practitioner do? The patient is too often a navvy, living in a doghole lodging, without a soul to attend to him, and it is nobody's business (except, of course, that of the unfortunate doctor who is first dragged out to see him) to furnish him with anything. Here is a great argument for the cottage hospital. Such an institution, with a few clean beds, a well-trained nurse, and a moderate collection of instruments kept in good order, is capable of saving an infinity of suffering. While, therefore, the profession is rightly beginning to discountenance the eternal multiplication of pettifogging special hospitals, got up by enterprising young doctors for their own benefit, and pushed by active committeemen for the sake of social position, no reasonable objection can be made to the erection in small manufacturing towns and villages of cottage hospitals for the treatment of urgent cases of injury. The one argument against them is that they may be diverted from their legitimate purpose, and made places where rival practitioners will make efforts to "get up their names" by attempts to extract kidneys or extirpate ovaries. It is sincerely to be hoped that such will not occur, and that they will be reserved for their legitimate function of rendering aid to injured persons. In this the medical men of every manufacturing town or mining village have that skill and experience which entitles them to deal with such cases, and that, too, as a rule, with great practical sagacity and shrewdness.

Partial Division of Radial Artery—Secondary Hæmorrhage.—One case—that of Ellen T., aged 9—may be alluded to as a warning to house surgeons. One Sunday the child, having been sent for some beer, fell and broke the bottle, which inflicted a deep wound on the front of the arm about midway between wrist and elbow. The bleeding was profuse; and she was at once taken to a public institution, where the wound was tightly bandaged up. The mother, with the careless negligence common to the poor, never took the child back till the following Thursday. The bandages being removed, severe bleeding commenced again. Once more tight bandaging. That night child vomiting and hand discoloured. Next morning (Friday) to the institution again, and arm re-bandaged, but no bleeding this time. Child in great pain—very cold, very sick, and slightly delirious. On Saturday to the institution once more, where the bandaging was abandoned and poultices ordered as the hand now looked black. The next day, a week after the accident, the child was brought to the Infirmary, with the arm gangrenous up to the wound, and the whole limb swollen and red up to the shoulder. Lead and opium lotions were applied to soothe the latter condition, and the miserable little patient was well fed and cleaned. Two days later the arm was amputated below the elbow and the child recovered. It was found that the radial artery had been cut across for about two-thirds of its circumference. At the site of the gaping orifice was a small false aneurism filled with clot. It is difficult to estimate the life-long injury done to this unfortunate child by what was either a piece of unpardonable negligence or gross ignorance—equally culpable in either case. The administration of an anæsthetic and the ligaturing of the vessel above and below the point of injury would not have taken twenty minutes, and would have saved the arm.

The mortality after amputation of the thigh and leg is always interesting, provided that a sufficient number of cases be taken to make the statistics serviceable. Small numbers are of no use. I happened by chance to stumble across four ponderous volumes the other day containing the statistics of all the hospitals of Paris, from

1861 to 1864 inclusive. It appeared that after amputation of the thigh in traumatic cases 77 per cent died; in pathological cases 57 per cent. After amputation of the leg 80 per cent. of the traumatic cases died and 40 per cent of the pathological. A Parisian hospital twenty years ago must have been a perfect slaughter-house. I looked up the strangulated hernias also out of curiosity. During the four years 227 cases were subjected to operation, of whom 172 died and only 55 recovered—about one recovery to three deaths!

As regards artificial limbs, the principle which Thomas has embodied in his knee-splint of bearing the weight of the body upon a pelvic ring encircling the upper part of the thigh, may very conveniently be applied to them. The drawing opposite page 16 represents the lad with the double amputation wearing a knee-splint as an artificial limb. Mr. Critchley, of Upper Pitt Street, Liverpool, has made a most cheap and serviceable pin leg for

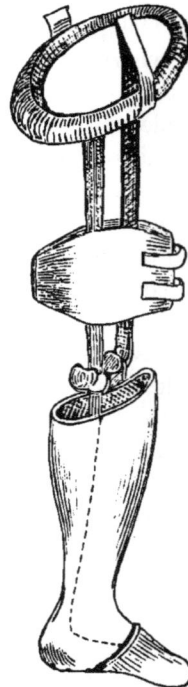

Fig. 1.   Fig. 2.

The pelvic ring principle of Thomas's knee-splint applied to artificial limbs.

working men, by uniting the two limbs of the knee-splint into a pin below. It is sold for a guinea, and for strength and lightness cannot be beaten. Even to more expensive forms of artificial limbs the principle can be adapted as is seen in figure 2, which is intended for an amputation below the knee-joint. The weight of the body is borne by the pelvic ring, an artificial joint corresponds with the knee flexure, and the fore-part of the foot is cut out of solid rubber, so as to give spring and elasticity without using complicating cords and pulleys.

EXCISIONS OF JOINTS.

TABLE OF EXCISIONS AND RESULTS.

Sex.	Age.	Disease.	Joint.	Result.	Cause of Death.
M.	12	Morbus coxæ,	Hip, ...	Died, ...	Exhaustion.
M.	40	Gelatinous disease, ...	Knee,...	Recovered.	... ...
F.	49	Do. ...	Knee,...	Died, ...	Exhaustion and bed-sores.
M.	52	Disease arising from injury,	Elbow,	Recovered.	
F.	12	Gelatinous disease, ...	Elbow,	Recovered.	
M.	10	Do. with destruction of cartilages,	Shoulder	Recovered.	... ...

In the few excision cases performed there were hardly any points of interest. The hip case was of two years duration, and, when the joint was excised, the head and neck of the femur were carious and much eroded, the acetabulum was perforated and the left pubic ramus was bare. The patient died from exhaustion between three and four weeks after the operation. In the two knee cases the disease apparently began spontaneously—at all events there was no history of injury. In consideration of the ages of the patients—40 and 49—amputation was recommended to them. This they refused, but consented to excision. The female died three and a half months after the operation from exhaustion and bed-sores. No union whatever had taken place between the bones. The male recovered after a prolonged struggle for life. He was six months in hospital. More than two years after he left he presented himself. The knee was sound, but nevertheless he had to use a pair of sticks, although with these he could walk for about two miles. Still he was of no use as a labourer and had never been able to do any work. Long sickness and privation had quite worn him out, and he looked as if he would soon go into a consumption. He bitterly regretted not having had his leg amputated.

As to the method of excising the knee-joint I have tried nearly every possible variety of incision, even including a semi-circular incision made above the patella, so as to divide the tendon of the quadriceps and the muscles above the joint in place of the ligamentum patellæ. The usual semi-lunar incision is undoubtedly the best, although every now and then somebody says he has found a better. The only real improvement which has recently been made in the operation consists in sawing the femur after a wedge-shape and making a notch in the tibia to receive it. Better still, the femur may be rounded off and the tibia hollowed out, a proceeding readily accomplished by the aid of a Butcher's saw. I have been doing this for the last two or three years with the best results in the way of steadying the limb and preventing the falling back of the femur, which is so apt to occur even with the best of splints. Dr. Fenwick, of Montreal, has been following out this practice very indefatigably for a long time. In the cases where the bones are not much diseased I have adopted a little expedient which is quite worth while trying. It consists in saving the posterior crucial ligament. This is attached very low down on the popliteal notch of the tibia posteriorly, and can certainly be saved in all the cases for old ankylosis and in most of those for disease. It helps, in conjunction with the bevelling of the bones just alluded to, to steady the parts. In three instances recently I have, by careful antiseptic treatment and carrying out of the above details, succeeded in healing the whole wound in a remarkably short space of time without a single drop of suppuration.

Never having seen any special advantage gained by retaining the patella, I always remove it in cases operated upon for disease,

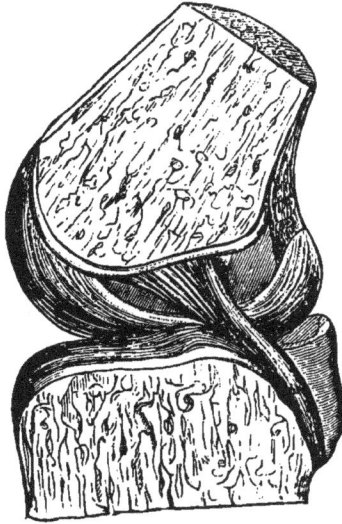

Vertical section of the knee-joint, shewing the attachment of the posterior crucial ligament low down on the back of the tibia.

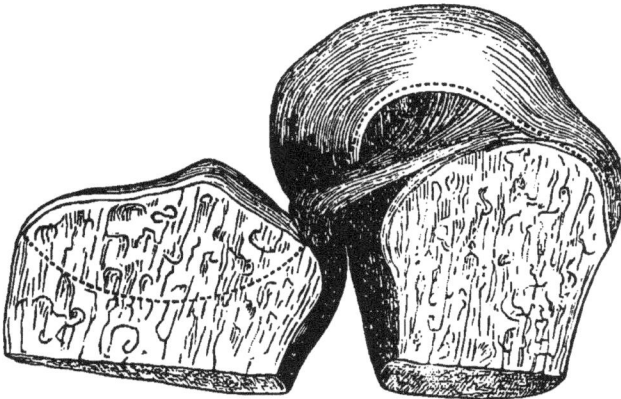

The dotted lines shew the amount of femur and tibia to be removed in the excision of the knee-joint in cases where "bevelling" can be practised along with retention of the posterior crucial ligament.

lest, by becoming carious, it should add another to the sufficiently numerous risks of the operation. The retaining or not of the patella used to be one of the great points of discussion in this operation. A far more important one was never thought about:—viz., the thorough removal of the diseased soft tissues of the joint. The whole of the jelly-like synovial membrane should be carefully cut or scraped away. The sinuses should be cleared of their putrid contents and the exuberant granulations everywhere scooped out with a Volkmann's spoon till clean tissues can be seen all around. Unless this is thoroughly done any attempt at antiseptic treatment must be a farce, and, if thoroughly done, the necrosis and subsequent removal by suppuration of the lowly organized debris will be avoided.

The progress of time and the multiplication of experience are gradually assigning to the various excisions their proper places in surgery. Of the excision of the shoulder and elbow-joint there can be but one opinion:—that they rank among the finest of modern surgical proceedings. Excisions of the ankle-joint on the other hand and the allied operations for the removal of individual tarsal bones are much less heard of than they used to be, and will, without doubt, become less and less frequent. There are three cogent reasons against them:—(1.) By the time the excision comes to be done, not one bone but many, not one joint but nearly all the tarsal joints are effected. The excision is usually, therefore, an imperfect affair at the best, which comes to an amputation at a subsequent period. (2.) Where a "cure" does result, the foot, in all the cases which I have seen, has been a lame and incompetent affair, obtained after months of treatment—weary months both to patient and surgeon. (3.) By Syme's amputation we obtain a complete and certain removal of the disease, a rapid recovery and a stump which is much to be preferred to a weakly, floppy foot.

As regards excision, then, the question of the day is with respect to the hip and knee. I regret that I have not collected the statistics of the hip and knee excisions which I have performed, and therefore, cannot give figures, but only the general impression which my own cases and those of my colleagues have produced upon me.

In a paper read at the last International Congress, M. Ollier of Lyons, so distinguished for his experience and researches in all matters relating to excision, made some very caustic remarks upon the " inutility of statistics, which depend upon a great number of "facts, collected from all sides and having nothing in common but

"the title under which they are published. These statistics already
"encumber science and are more of a hindrance than really useful.
"Doubtless one must not cast them aside, for in spite of their in-
"conveniences they possess a certain degree of utility; but they
"ought to be accepted *sous benefice d'inventaire.*" And in truth what
possible comparison can there be between the statistics of him who
excises wholesale the joints of young children in an early state of
disease and the statistics of him who reserves his excisions as a
last resource? I daresay that during the last three years I might
have excised fifty hips and knees had I been desirous of doing so,
and, no doubt, could have brought out some excellent statistics as
an early exciser, to have flung in the teeth of the late exciser.
But would that be surgery? At the Congress there was great
debating about this same doing of early excisions; but of all the
speakers, Mr. Howard Marsh alone seemed to me to strike the
key-note of the whole question. He ignored statistics, and looked
to the broad surgical bearings of the subject. He pointed out that,
after all, excision belongs to the same class of treatment as amputation. It is a giving up the attempt to cure the disease. The
surgeon in fact throws up the sponge. Now a joiner after six
lessons will cut off a leg,—that is only surgical carpentry at the best;—
and the most indifferent surgeon will very soon excise joints as
well as the most expert. But it takes a clever surgeon to save a limb.
This is far higher surgery and far more important surgery; seeing
that we know perfectly what excision can do, and have no further
knowledge of moment to acquire concerning its mode of performance or subsequent treatment. The question, then, being not how
to do it but when to do it, I must say that an indiscriminating
advocacy of early excision seems to me a very dangerous thing
indeed in many ways, and notably dangerous to the progress of
true surgery. Far better for us to devote our energies to teaching
the early recognition of hip and knee disease, and to elaborating an
efficient treatment of their incipient stages. The children of rich
people don't have their hips and knees excised. Why not? because
the articular mischief is promptly found out, and skilfully and
patiently treated. That the children of the poor can ever be put
in the same position is of course out of the question, but every year
we are learning better how to manage their joints and how to give
them cheap and efficient mechanical rest, even at their own homes

I am quite prepared to admit the recoveries after early excisions, but what I assert is that these excised joints should and could be saved, if efficient treatment were soon enough commenced. Up to fifteen years of age the knee and hip can be almost certainly cured, if the disease be taken at the first blush. But too often the golden opportunity is lost through inability on the part of medical men to recognize the initial symptoms and promptly act upon them even at the risk of being thought alarmists. The early lamenesses of young children are continually being pooh-poohed as growing pains or bad habits, and not deemed worthy of having anything done for them. It is the appreciation of the awkward gait and the trifling limp that the medical student must be more carefully taught, and taught to regard as of the profoundest import. Then is the time for treatment— then literally is the day of salvation for the joint;—not when there are night startings and pain, and screaming upon movement. In performing and advocating these early excisions surgeons have got hold of the wrong end of the stick. Let us begin at the beginning, and when we know all about means of prevention and early treatment and have come to an end of that part of the subject, then we may talk about early excisions. Meanwhile we are only opening our eyes to the treatment of joints, and the past twenty years have done more than the previous three centuries. By the way, the report of the committee of the Clinical Society on excision of the hip-joint, presented in May, 1881, was not very encouraging. It made out a mortality of 35.5 per cent. in cases of excision as against 30.4 in cases of suppuration, treated by rest and extension. Furthermore, it was admitted that the limb that remains after cure by the latter means, although stiffer, is more firm and useful for purposes of progression than that left after excision.

From what I have seen of hip and knee cases my teaching to students is this:—In children up to fifteen years of age, if you get a case of knee or hip disease from its commencement, make up your mind to save the limb. You ought to save it. Between fifteen and twenty-five, failure is to be looked for very often, and then you may excise. Don't operate until your art is exhausted— don't delay until your patient is exhausted. Fortunately after twenty-five or thirty, joint mischief is not common; but at that age whatever you may do with the hip, do not excise the knee, if your patient will let you amputate.

I may have been unfortunate in my own practice and in that which I have seen, but the general impression left upon my mind is, that excision of the knee-joint after thirty years of age is, as a rule, a disastrous thing, and that many a life has been lost to save a leg. Besides, let us take the case of a working-man. I will be bound to say he does not do a stroke of work on an excised knee-joint under eighteen months to two years. After amputation he is at work in four to six months on a sound stump; and, while I do not wish to advocate too strongly one rule for the rich and another for the poor, I am convinced that if a working man, who is offered his choice of excision or amputation, could thoroughly forsee what awaits him with the former, he would choose the latter almost invariably. As it is, all he sees is losing or keeping his leg—a point to which some of them attach a greater importance than to life itself.

TABLE OF FRACTURES AND DISLOCATIONS.

SIMPLE FRACTURES.

Nature.	Number.	Recovered.	Died.
Nasal Bones,	1	1	—
Lower Jaw,	1	1	—
Do. and radius,	1	1	—
Neck of scapula,	1	1	—
Humerus,	1	1	—
Do. ununited—operation,	1	1	—
Do. and both clavicles,	1	1	—
Radius,	1	1	—
Do. and ulna (refracture),	1	1	—
Ribs,	7	7	—
Do. with resection,	1	1	—
Femur,	7	7	—
Tibia,	15	15	—
Fibula (Pott's),	2	2	—
Tibia and fibula,	10	10	—
Patella,	5	5	—
Spine,	2	0	2

COMPOUND FRACTURES.

Nature.	Number.	Recovered.	Died.
Upper and Lower Jaws,	1	0	1
Humerus into Elbow Joint,	1	1	—
Femur,	4	2	2
Tibia (complicated with simple fractures elsewhere),	2	2	—
Tibia and fibula,	6	5	1

SIMPLE DISLOCATIONS.

Nature.	Number.	Recovered.	Died.
Forefinger (excision),	1	1	—
Humerus—irreducible	1	1	—

COMPOUND DISLOCATIONS.

Nature.	Number.	Recovered.	Died.
Elbow Joint,	2	2	—
Great Toe,	1	1	—

Fracture of the Neck of the Scapula.—Among the fractures of the bones of the upper extremity an interesting case was that of Catherine S., an old woman of 74, who was knocked down and run over by an empty coal cart. On admission she was suffering from shock and complained of her right shoulder. On the following day, having recovered from her condition of collapse, the site of injury was carefully examined. The skin over the back of the right scapula was bruised and movement of the shoulder gave pain. The shoulders and scapulæ of opposite sides being compared, no apparent difference could be found between them. The right acromion and the spine of the scapula were then traced and found quite intact. The whole scapula was then steadied, and the arm, being rotated, was seen to move easily and painlessly at the shoulder-joint, shewing that there was no fracture of the neck of the humerus. Nevertheless, when the patient herself voluntarily moved the shoulder there was pain, and during our various manipulations a frequent and distinct sound of crepitation was heard. The blade of the scapula was therefore carefully fixed, and then, the shoulder as a whole being grasped and moved up and down, crepitus was at once produced. This experiment was repeated again and again till no doubt remained that there was a fracture somewhere between the glenoid cavity and the blade of the bone. This fracture was of necessity external to the root of the spine, otherwise that prominence would have been broken also, which was not the case. As it is believed that fracture of the anatomical neck—the nar-

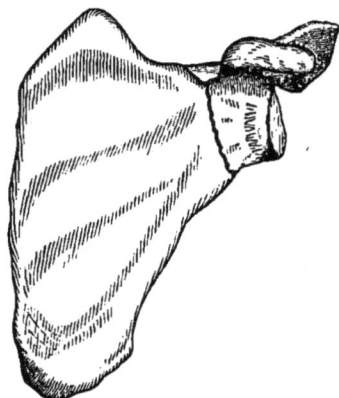

Fracture of the surgical neck of the scapula.

row part immediately beyond the glenoid area—is practically impossible, or it all events has never been seen, it follows that, in the present instance, the fracture must have been the one described by Sir Astley Cooper, which, commencing above at the supra-scapular notch, cuts off the coronoid process, glenoid cavity and anatomical neck from the blade, leaving the acromion and spine intact. In making this diagnosis the two chief reasons were (1.) that when the whole scapula and glenoid were fixed, the humerus moved freely in the glenoid area and no crepitus was audible, and (2.) that when the blade of the scapula alone was fixed and the shoulder moved up and down, crepitus was immediately and always produced. In "Hamilton's Fractures" the signs of this accident are given; notably a falling down of the humerus and attendant fragment of the scapula, which is readily done away with by support, but recurs as soon as the support is taken away. But it is admitted that, if the coraco-acromial and coraco-clavicular ligaments are not ruptured, very little displacement of the fragments may occur. Such was very likely to have been the case here, looking at the manner in which the accident occurred. For it is not difficult to conceive that, the old woman having been knocked flat on her face, the wheel of the cart, as it mounted over her shoulder and back, may have so pressed upon the upper part of the humerus and neck of the scapula, as to break the latter off from the rest of the bone, and yet leave the various ligaments attached to the coracoid process uninjured. If this view is correct, the injury was a very rare one, and being the only fracture of the scapula that has as yet come under my observation, more than ordinary pains were taken with the examination. The arm was suspended in a sling and kept bandaged to the side and in a few days the patient left the Infirmary and attended the out-door department. When last seen, about seven weeks after the accident, she could move the shoulder without pain, but pressure about the neck of the scapula still hurt her.

Ununited Fracture of the Humerus—Two Resections.—The patient was a very powerful, ruddy, healthy looking collier, thirty-two years of age. Ten months before admission he fell and sustained a simple fracture of the right humerus in the middle third. A mere fibrous union resulted and the arm was practically useless. Under antiseptics the ends of the fractured bone were sawed off and the fragments retained in position by a large leather splint which fixed the shoulder and elbow. Twelve days afterwards the wound was

practically healed and the arm was rendered absolutely immovable by a hoop iron splint let into a plaster casing, which enveloped
the upper part of the chest, the shoulder and the arm down to the
wrist.   Not the faintest movement could occur.   For many months
the arm was kept thoroughly steady, but in spite of this no proper
union occurred.   Phosphorus internally was tried on the strength
of its being said to produce bony growth in dogs.   Next we tied a
handkerchief tightly round the arm above the fracture, so as to
keep a large supply of blood in and around the callus;—to congest
it in fact.   Finally three steel drills were driven through the callus
and left for about three weeks until they dropped out of their own
accord—still no union.   So, seventeen months after the first operation, the old incision was re-opened and the old condition discovered again, namely, a most perfect fibrous union.   The fibrous
mass even contained a cavity which held a little fluid.   The bones
were sawed obliquely and an ivory peg was driven through both of
them at one part and a strong silver wire at another—holes having
been bored with a dentist's drilling engine.   On this occasion antiseptics were not employed.   During the subsequent thirty-six hours
there was so much oozing that the wound had to be opened and
stuffed with lint soaked in turpentine.   After that all sorts of
abscesses and suppurations formed, resulting in the speedy carrying
away both of ivory peg and silver wire.   The arm was simply hung
in a sling and poulticed and for a month the patient was very ill
indeed.   By degrees the wound granulated up and an inside rect-
angular splint was got on.   At the end of about four months there
was complete union with an enormous mass of callus.   It was a
great question with us whether the primary union obtained at the
first operation, combined with the absolute immobility of the fragments, so far from being serviceable were not positively conducive
to non-union.   In fact there was not sufficient irritation to induce
the formation of ossific material around the recently sawed ends of
the bone.   That there is some truth in this idea is shewn by the
fact that after the second operation (which was conducted without
antiseptics) there was such an amount of inflammation and suppuration that the arm could not be kept in any splints at all, and the
fragments were absolutely unsupported and moved about anyhow.
Nevertheless, an abundant callus formed and excellent union resulted.   If a fracture seems very slow to unite, the orthodox treat-

ment is to apply the splints more firmly and enjoin perfect rest more rigidly than ever. I have began to entertain strong doubts about this policy, and under such circumstances have several times taken all splints off and allowed the fracture to take care of itself and get some knocking about, with the result of finding a firm union take place. After all, this is simply a minor form of that artificial imitation, which at a later period we seek to produce in a variety of ways, if the fragments still refuse to unite. Another point to be noticed is that, if the non-union is due to the fact that a stout piece of muscle or fascia has got between the ends of the bones and so prevents their direct contact, prolonged rest can do no good at all and movement may,—who knows how often this is the case?

To revert to the subsequent history of the case, the patient left the Infirmary with the humerus united by such a huge mass of callus as I never saw in any other case. Coming back after a while he complained that while the arm in other respects had recovered its strength he could not extend his fingers. Flexion was strong enough so that he could grip sufficiently well, but as he could not open the hand again the fore-arm was extremely useless. It became pretty clear that something had happened to the musculo-spinal nerve; either it had been divided at the operation, or else, when the bones subsequently fell apart and the great inflammation and suppuration occurred, it had got mixed up in the enormous mass of callus that resulted and so become squeezed. As against the idea of its being divided at the second operation it was urged that it certainly was not divided at the first one, and at the second the old scar was simply re-opened. On both occasions great care was taken to look out for it.—In the end of 1883 the patient came to see if anything could be done, as, although the arm possessed a certain amount of usefulness, the loss of extension power in the fingers prevented his following his occupation as miner. The humerus was splendidly firm. The callus had much diminished in amount, but there was still a great mass of it. Under strict antiseptics I cut down upon the musculo-spiral nerve, as it lies between the pronator teres and the brachialis anticus and followed it up. It was found to run into a mass of fibrous tissue and soft callus—half cartilage, half bone in appearance—in which it gradually disappeared. Above the callus it was once more discovered, but there was far too great an interval to permit of any attempt at union being made—it was quite impossible to trace the deficient piece.   That the nerve had been swallowed up in the inflammatory products out of which arose the great mass of callus seemed hardly doubtful.   Had it been divided at the second operation its cut ends would have been tolerably well defined and would have been found near each other, seeing that as the upper arm was shortened and not lengthened by the operation, there could have been no strain upon the nerve.   The wound and dissection necessary to clear up the mystery about the nerve were very extensive, but immediate healing ensued and the patient was rapidly *in statu quo* :—*i. e.*, with a useful upper arm and loss of extension in the fore-arm.   A disappointing conclusion to what promised so well.

Colles's Fracture.—Colles's fracture seldom presents anything out of the common, but the case of M. R. had some features of interest.   He was a poor, broken down, rheumatic creature, sixty-five years of age.   Three months before admission he sustained a Colles's fracture of his right radius and was for some weeks an inmate of a workhouse hospital, where the fracture was treated, and from which he was discharged when it seemed to be duly united. Nevertheless he found himself unable to use the hand from stiffness of the wrist and fingers, and so came to the Infirmary.   The chief point was that the hand, as presented to us, displayed in the most characteristic way the deformity of a Colles's fracture which had been only recently sustained and which had never been reduced.   The dresser's notes say " The deformity is well marked, " the hand and wrist being carried backwards and producing an " elevation posteriorly, while the flexor tendons are thrust forward " in front."   A careful examination shewed, however, that the fragments were really in position and apparently united, and that the deformity was in the soft parts.   The arm was placed in splints and kept at absolute rest for a month, at the end of which time the deformity had quite disappeared.   The case was a most useful one clinically for students, and the condition corresponded exactly with Hamilton's description of it as " a broad, firm, uniform swelling on " the palmar surface of the fore-arm, commencing near the upper " margin of the annular ligament and extending upwards two " inches or more.   This swelling continues much longer in old and " feeble persons, than in the young and vigorous.   It is pretty

"generally proportioned to the amount of anchylosis existing at
"the wrist and finger joints, and it disappears usually *pari passu*
"with these conditions.   There can be no doubt that this pheno-
"menon is due to effusion along the sheaths of the tendons and in
"the areolar tissue external to the sheaths, and it is as often present
"after sprains and other severe injuries about this part as in frac-
"tures.   In many cases, however, its prolonged continuance and its
"firmness have led to a suspicion that the bones were displaced, a
"suspicion which only a moderate degree of care in the examination
"ought easily to dispel.   A similiar effusion but in less amount is
"frequently seen also on the back of the hand below the annular
"ligament.   When both exist simultaneously the appearances of
"deformity and displacement are greatly increased."

A few words on the treatment of Colles's fracture may be
pardoned.   Just as the obstetrician sooner or later invents a new
midwifery forceps, or the gynæcologist an improved speculum, so it
seems to be incumbent on the surgeon who interests himself in
fractures to manufacture a patent splint for the injury under consideration.   So that the apparatuses for this purpose are innumerable.   For my own part I have never seen the slightest occasion
for anything but two flat, narrow, well padded splints, such as
Colles himself originally recommended.   But it is of the utmost
importance *that the dorsal one should not extend beyond the knuckles
nor the palmar one beyond the ball of the thumb.*   By means of the
bandage the hand can be drawn inwards somewhat so as to assume
the pistol shape, if desired, but the thumb and the fingers should
be left quite free.   By this means the hand is ready for use as soon
as the splints are removed, without any of the dreadful stiffness
which one too often sees after the use of splints which confine the
fingers.   The movement of the fingers does not tend in the least to
displace the fragments.   In fact if a thorough reduction of the
fragments has been made at first there is no tendency to displacement at all.   But this thorough reduction should be made as soon
as the fracture is seen and made once for all.   To that end, if there
is much outcry and resistance to manipulation on the part of the
patient, an anæsthetic is greatly to be recommended.   Specially
is this desirable in private practice so that the reduction may be
completely and indubitably effected, to the prevention of all subse-

quent recriminations, which are perhaps more frequent with this than with any other fracture.*

The exact nature of the injury to the bones and soft parts in Colles's fracture has given rise to much writing, and there has been great difference of opinion as to whether impaction of the fragments as a rule does or does not exist. I am firmly of opinion that it does exist, and that the undoing of this impaction is the essential feature in reduction. Solemn warnings are uttered in text books about some imaginary evil which may befal old people if the impaction is undone, but having undone it over and over again, with the result of avoiding all deformity and without encountering any evil consequences, I have ceased to respect these warnings. To my mind there is one crucial proof of impaction. Given a non-impacted fracture of any long bone in the body, you can always find some position of the limb in which, all the muscles being relaxed, the fragments can be induced to lie in perfect line— the sole difficulty in treatment is to keep them so. But in Colles's fracture you may put the arm in any position conceivable and you won't alter the deformity one bit: you won't alter it until by the use of more or less force you have unlocked the bones. Having the patient's arm firmly fixed just above the fracture, I follow the usual practice of taking his hand in mine, as if to shake hands, and then making steady and forcible extension. The hand is next bent well down to the ulnar side (adducted) and the lower fragment pushed forward into its proper place by firm pressure on its posterior surface. Curiously enough, Hamilton severely reprobates the practice of extension at the wrist, as "adding to the fracture "and to the other injuries already received the graver pathological "lesion of a stretching, a sprain of all the ligaments connected "with the joint. I am persuaded that to this violence added to "the unequal and too firm pressure of the splints, are in a great "measure to be attributed the subsequent inflammation and "anchylosis in very many cases." This is very dreadful to read, but many eminent men who write books write theories, which small men, who only do practice, find out to be theories and

* It is satisfactory to find that so excellent an authority as Dr. Packard in the article on Injuries of Bones in Ashurst's Encyclopædia of Surgery, considers that the great reason why deformity often follows Colles's fracture is, that reduction is not effected at all, and, moreover, is of opinion that it is important to act as much as possible on the fragments themselves. This has long been my own conviction, although I have hesitated to express it.

nothing more. To produce the slightest injurious effect on the ligaments of the wrist joint, supported as they generally are by the tendons of muscles in full contraction, would require infinitely more force than any one man could effect, provided always that the force be applied steadily and in a line with the arm. Sprains are never produced in this way: they are produced by sudden jerks out of the usual axis of motion of the joint.

I constantly endeavour to impress upon students three things about a Colles's fracture. (1.) Put the fragments in absolutely perfect position at the beginning and they will stay there. (2.) Let the splints end at the knuckles on one side and ball of the thumb on the other, leaving the thumb and fingers free. (3.) Let those be gently moved all through the treatment from the very first day. By adopting these rules I contend that deformity and stiffness need hardly ever be seen.

Fractures of the Ribs.—Note on Empyema and Removal of Portions of the Ribs.—Eight cases of fractured ribs were sufficiently severe to require treatment in the wards. Of these the case of Sam Fraser interested us very much. He was a sturdy, healthy young fellow, twenty-two years of age, who got his chest crushed between the buffers of two railway waggons. He was unconscious when brought to the Infirmary. His left clavicle was broken, and also four or five ribs on the same side, from the third to the eighth, near their anterior angles. By the sharp end of one of these broken ribs the lung was evidently perforated, as he spat blood, and a subcutaneous emphysema of the left chest speedily came on. At the same time the lung collapsed, and the left pleural cavity became filled with air. For some days he was in great danger from dyspnœa and exhaustion, but gradually rallied. The surface emphysema disappeared, but internally, in place of air, we found the pleural cavity filling up with fluid. About the third week after admission dyspnœa became again very urgent, accompanied by pyrexia, so that it became necessary to tap the chest. It was then found, as was suspected, that the fluid had become purulent, so that he had gone through the stages of pneumothorax, hydrothorax and empyema. The pus of course re-accumulated, and there was another tapping, and another re-accumulation, so that it was decided to open the pleural cavity and drain it. Two openings were made under antiseptics in the lateral line of the thorax, one through the fifth intercostal space and the other through the space just above the attached margin of the diaphragm. The antiseptic dressing was maintained for about six weeks, at the end of which time we resorted to simple washing out and the use of absorbent wadding to soak up discharge. During the ensuing six months the patient made slow but steady progress, only interrupted once or twice by feverish attacks when the tubes got blocked up and some retention took place. At last things came to a standstill. His lung had expanded as far as it would, his heart had come over to its nominal position as far as it would, his diaphragm had come up as far as it would, and his ribs had come so close together that there were no intercostal spaces left. And still there was a considerable cavity enclosed by walls which would not "give" any further. I had just heard from a Belgian surgeon of the removal of portions of rib to allow of a greater falling in of the chest wall in cases of this kind. It was before this proceeding had begun to be much, if at all, noticed in British Journals. Accordingly I cut down upon the sixth rib, opened its periosteal sheath and excised a piece about two and a half inches long, and a short time afterwards, I cut out about three inches of the fifth rib immediately above the other. There was some little difficulty in fishing out the first piece, because the ribs had come so tightly together that a knife blade could not be forced in between them. The patient's left chest wall, was in fact an impenetrable bony cuirass, through two holes in which the elastic drain tubes passed. They had successfully resisted the compression of the closing ribs and maintained patent the apertures in which they lay.

What good came of removing these portions of ribs? None that I saw. There was no remarkable or special collapse of the chest wall as a result, and in the course of a few months the periosteum had secreted bone enough completely to fill up the gaps. After a long period of slow convalescence the patient became well enough to obtain light work as a railway pointsman. By degrees the cavity greatly filled up, doubtless by the organisation of granulation tissue. He steadily worked for more than two years, and when last seen the cavity seemed to have contracted down to a sinus about four inches long, in which he wore a drain tube, and from which came about a drachm of pus in the twenty-four hours. One of the two original apertures quite closed up. At the apex of the left lung and beside the vertebral column breath sounds were audible; but it was plain that the left lung must have been very early bound down and never expanded, so that he breathed almost entirely with his right. His general health was very good, but he never quite recovered his former strength, although he was able to do a fair day's work. Only the other day I heard by chance of his somewhat sudden death, from some acute brain trouble, about three years from the time of the opening of the chest. He was still wearing the tube.

With regard to the removal of portions of rib in cases such as that just narrated, I do not say of what value the operation may be in children, but I am convinced that in adults it is absolutely useless. The ribs become jammed together until the chest wall is a plate of bone. In this you cut a window—that is all. And the pleura and sub-pleural tissues are so thickened and condensed that even over this area they cannot fall in at all. So that even the little advantage that might have been expected from this is not realized. It is obvious that to do any real good a couple of ribs— say the fifth and sixth—would have to be removed in their entirety from sternum to spine. Then, no doubt, the fourth rib would come down upon the seventh, the chest wall would fall in and the spinal column would bend laterally. But anything short of this seems absolutely useless, so that I shall not practice partial removal again, and complete removal is too serious a proceeding to undertake, in view of any good to be derived from it. It has been asserted that, after the insertion of the drainage tubes and emptying of the chest, the coming together of the ribs will nip the drain tubes and close them, and that, therefore, pieces of rib should be removed to prevent this. That this idea is purely theoretical, is shewn by the fact that in the present case the ribs of a full grown man closed down as tightly as they possibly could on two rubber tubes, but so far from these becoming nipped, their steady pressure ate away the ribs, on each of which there was a half circle that looked as if cut cleanly out with a drill or a gouge.

Fractures of the Femur.—Seven simple fractures of the femur were admitted, of which six were treated on Thomas's knee-splint, a method of practice which is as yet almost peculiar to Liverpool. It is employed as follows:—carry a long strip of plaster from the knee down one side of the leg and up the other leaving a loop or stirrup opposite the sole of the foot. This strap must be fixed to the leg by cross straps and held in place by a neatly applied roller that it may not slip in the least. Just within the stirrup a flat piece of wood, about two and a half inches from side to side by an inch from front to back, should be inserted, having a hole in its centre. Through this pass a strong rubber cord, knotted above to prevent it slipping through the hole. The use of this small piece of wood is important, as it prevents the stirrup strap from pressing on the malleoli, an occurrence which often gives the patient a great deal of pain. A Thomas's knee-splint for bed use should now be selected, fitting well above and reaching about six inches below the sole of the foot. The splint being adapted the leg should be steadily hauled downwards, so as to pull the fragments of the femur into line. While an assistant retains the limb in the extended state, the ring of the splint should be well pressed up against the pelvis, and the rubber cord should be made tense and fastened to the cross bar at the foot of the splint. By this means a steady elastic extension is kept up, the counter-extension being made by the pressure of the splint ring against the pelvis. Two short splints, made of Gooch splinting, should next be obtained. They should reach the whole length of the thigh and should be applied one in front, the other behind. In the Infirmary there is kept a number of such splints in stock, only made of tin and lined with boiler felt stitched to them through small holes along the edge. They are of various lengths to suit various thighs. These two short splints should be lashed to the thigh with a couple of webbing straps and buckles. Their object is to steady the fragments in the immediate neighbourhood of the fracture. Finally the leg and the thigh should be bandaged to the Thomas's splint, so as thoroughly to fix the whole limb. If the splint is a well fitting one and the above mentioned details have been carefully carried out, the fracture is so firmly fixed that you may take hold of the lower end of the splint and swing the limb about at the hip-joint without giving the patient pain. I have often done this to shew students and strangers how completely the limb is at once extended and rendered thoroughly immovable. Points to be observed:—(1.) See that the ring does not cut the patient on the inner side of the thigh—any cutting or chafing here upsets the whole treatment as it prevents the extension being applied.

(2.) Let the patient lie quite flat on his back. The temptation to sit up is very great; but if this is allowed he will, in the act of pulling himself into the erect position, infallibly tip out of place the upper fragment—no splint in the world will prevent this. (3.) Let the bandaging of the thigh be from time to time undone, and let the short splints be tightened up over the fractured ends of the bone, noting at the same time that these are keeping in good position.

After a trial of all the usual apparatuses for fractured femur, I now use the Thomas for all fractures in the middle and lower thirds till close upon the condyles. I have found no method *more*

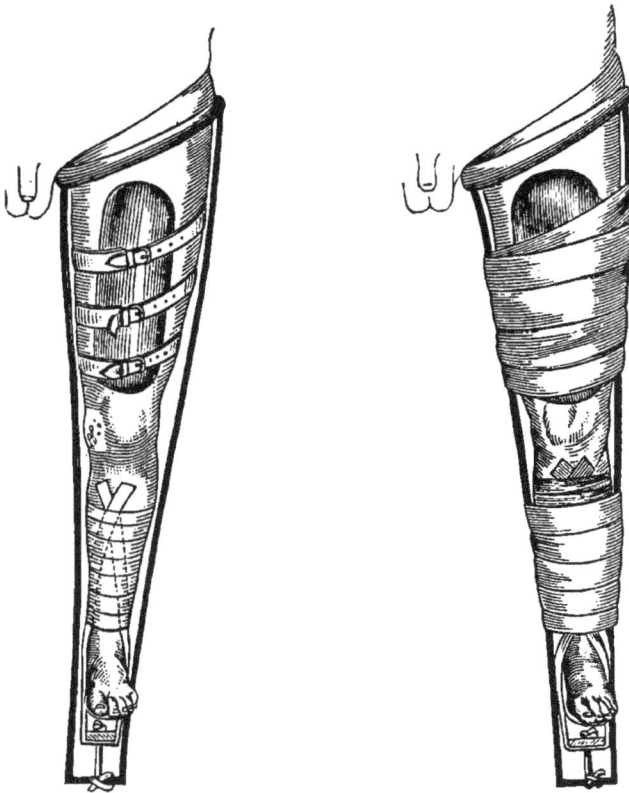

Fractured femur treated on Thomas's knee-splint.

satisfactory in its after results as regards shortening, and there is certainly none so comfortable to the patient during the treatment.

It allows all the ordinary movements for bedmaking and cleaning to be performed without pain, and gives the patient just that little power of easing himself from side to side, which often makes the difference between comfort and misery, and which the long splint does not permit. John Bell said, " The machine is not invented by which a fractured thigh bone can be perfectly secured." I believe he would admit it was now, if he saw one well put up in a Thomas's splint.

For fractures in the upper third I prefer the M'Intyre—or double inclined plane; because the tendency to tilting up of the upper fragment by the psoas and iliacus is sometimes so great that it is best to bring the lower fragment up to it. I am not speaking by the book : I have seen this. Again, in fracture close to the knee, the M'Intyre is also the best, on account of the tendency of the lower fragment to fall backwards. Here it must again be brought up to the upper one, although for a different reason. Hamilton says—" There are a few, however, of our most distinguished surgeons who retain the flexed position in certain fractures, such as an oblique downward and forward fracture, occurring just below the trochanter minor and a similar fracture just above the condyles." To the practice of this eminent, if not numerous body, I venture respectfully to give my adhesion.

Fracture of the Patella.—Five cases of this accident were treated. In watching these cases, one is struck with the rapidity with which the joint often fills with fluid after the injury, and with the amount of tension which ensues. This fluid is, doubtless, almost entirely blood. I had no idea how much blood could be effused from the fractured surfaces of a patella until I happened to examine the joint of a patient who had just been brought in with a compound fracture of the bone. The blood literally poured from the broken ends. There is no reason why the fluid effused after a fracture should be anything but blood. Simple fracture is not produced by a blow or by direct violence, or then there might be inflammatory serous effusion, but the patella is merely snapped in two over the trochlea by muscular action—sometimes not very excessive. Should the amount poured out be moderate, and allow a reasonable approximation of the fragments, one, of course, trusts to time for its absorption. But if it is excessive—rendering the capsule tense, and preventing the fragments coming together—

there need be no hesitation in tapping the joint. I have done it several times, with the effect of saving a good deal of time and pain.

One of the cases was a serious one. Robert M'Gill, a seaman, aged 30, fell on board ship in the Gulf of Mexico, and fractured his right patella. From his own statement it was clear that the treatment had been very ineffective. He came to England in a loose, short, plaster of Paris case; and what little union might have been taking place under this was ruined by his falling on deck while the ship was rolling, and tearing everything apart. He came to the Infirmary in the beginning of March, 1881. The joint was found much distended with fluid, and the fragments widely apart. The upper one was much the larger of the two, the lower being only about a fifth of the bone. This, doubtless, contributed to the great drawing up of the upper fragment. On March 4th the joint was tapped, and about four ounces of bloody serum were drawn off. Four days afterwards it was again tapped, and about two ounces removed. By the 12th it was quite free from fluid, and the fragments were much nearer each other; but still they would not come as close together as was desired. Having read in the *Medical Press* for February, 1881, a notice of an operation by Kocher, I resolved to try his plan; so, on the 14th, under ether and antiseptics, the fragments were brought together as follows:—A sailmaker's large "palming" needle was procured, and armed with very thick silver wire. It was thrust into the joint through the quadriceps tendon above the upper margin of the patella. Being then pushed beneath the fragments, it was brought out inferiorly through the ligamentum patellæ. The fragments were then pressed together, and the ends of the wire tightened

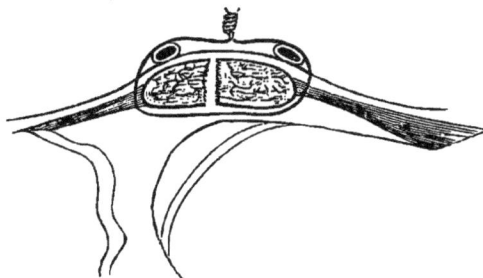

Section of knee-joint. Patellar fragments brought together by a wire passed round them. The wire prevented from cutting the skin by pieces of rubber tubing interposed.

over the outside of the patella, while two pieces of rubber tubing were placed beneath it to prevent its cutting the skin. The leg was then placed in a Thomas's knee-splint. The wire was retained in position for twenty-five days, all the time under antiseptic dressings. At first the patient complained of a good deal of pain but there was never the least sign of any inflammation in the joint. The subsequent treatment consisted in strapping the limb and maintaining it straight on the splint, which was continued till the end of July, (three months and a half from the time of the operation,) when the patient was dismissed with the fragments in excellent position. Most unfortunately I was away on my annual holiday when he left and the result was that he was not enjoined to keep on wearing the splint, and so left it off. Having no support to the still weak limb he was hardly a fortnight out of hospital, when he fell down on the street and tore the fragments apart as widely as ever. As I was away, he preferred going to the work-house hospital to returning to the Infirmary; and went under the care of my friend Dr. Alexander, who opened the joint, refreshed the ends of the bones, and brought them together with silver wire. He recovered with a stiff knee it is true, but with a very useful limb. It was most annoying that the patient's carelessness should have undone what was effected in the Infirmary with considerable trouble, as the plan of wiring seemed one excellently adapted to such a case. Where a long thin fibrous band exists between the fragments, clearly the only thing is to open the joint, remove this and refresh the ends of the bones. But in this case no such fibrous band seemed to have formed and the non-union and separation were due to the fact, that the patient kept moving about after the accident for some weeks while the joint was full of bloody serum.

Everybody seems to have his own plan of treating a fractured patella; some using apparatus of the simplest, some of the most complicated description. The more I see of these fractures the more am I convinced that, so long as the patient lies flat in bed, it is a matter of indifference what apparatus is used. Some straps of sticking plaster well applied will do all that is really necessary in keeping the fragments together. As a matter of precaution it is no doubt a wise thing to put on a back splint, but even this is not an absolute necessity if the patient will lie still. Concerning the placing of the limb on an inclined plane, that also is a prudent measure; one that cannot do any harm and may do good, but not essential. Let the experiment be tried. After the patient's muscles have been quieted with a couple of days rest in bed, lay the limb flat on the mattress, and place a finger in the line of fracture. Then steadily raise the limb by the heel. The fragments will be found to remain perfectly still. The truth is that so long as the patient lies quite flat, the quadriceps extensor is perfectly relaxed. Moreover its lateral portions the vasti, which are inserted into the fascial and fibrous tissues at the side of the knee-joint, being seldom torn, tend to hold the fragments together. It is only when the patient raises himself into the sitting posture that he draws up the superior fragment, and this he does through the agency of the rectus which is inserted superiorly into the pelvis. In all but the worst cases the fragments can be brought within a quarter to a third of an inch of each other and maintained so. Why then do we so frequently see such bad results, such weak legs, such intervals of one or two inches between the broken ends? Simply because the surgeon, having kept his patient in bed or on the sofa for six weeks or two months, commits the profound mistake of thinking that the tissue uniting the fragments is firmly consolidated and proceeds to bend the joint, in other words deliberately to stretch this soft extensile medium. "A very good mend," he says to the patient. "Only a third of an inch of interval. Now you may get up and move about and get the stiffness out of your knee." The patient does so; the soft ligamentous union yields, and both he and the surgeon are very much disappointed when, after a few weeks of going about, the interval is increased to an inch or two. This all arises from an unfounded fear lest a stiff joint should result from want of movement. If there has been no inflammation in it, or in the tendons or fasciae round it, a joint will remain unmoved for months and months, and eventually become perfectly supple again. At the end of a couple of months the patient, when he gets up, in place of being told to bend his knee, should be fitted accurately with a Thomas's calliper walking splint, which he should wear for six months more, and never bend his knee all the time. He will then have as good a leg as ever. Possibly most surgeons know and will admit this; but they certainly do not carry it into practice,

and in hospital work it is the usual thing to send a man
out at the end of six or eight weeks as " cured," when his cure
is really only commencing.   As a result of prolonged immobility of the joint, and non-use of the quadriceps, I have
seen two cases of undoubted bony union within the last
three years.   One of these was one of the five cases under consideration.   The patient was a labourer aged 35.   He fractured
his patella in the beginning of June, 1881.   After his discharge
he wore a Thomas's calliper splint, working on it as a quarryman, until the end of January, 1882.   When seen at that
time there was no appreciable hiatus between the fragments,
which were clearly united by bone.   Some years ago I treated
in private a young gentleman who broke first one patella and
then, twelve months afterwards, the other.   He was a noted
horseman and waltzer.   By keeping the knees motionless for long
periods he recovered with about a quarter of an inch of interval
between the fragments, and rode and danced as well as ever, never
feeling, as he often assured me, that anything had ever happened
to him.   Surely what can be done with one patient can be done
with another.   I cannot resist making these remarks in view of
the great interest which has been excited by Sir Joseph Lister's
paper, recently delivered before the Medical Society of London, in
which he narrated a series of admirable cases where, in consequence of bad union and uselessness of the limb, he had opened
the knee-joint, removed the imperfect band of union, freshened the
edges of the patella, and united them by suture, leaving a thoroughly useful limb.   One cannot but regard this as a triumph of
surgical art and of the antiseptic system, when it is remembered
that, not a quarter of a century ago, the mere fact of a joint being
opened was considered as bringing amputation within measurable
distance.   But one could not but be astounded at the opinion advanced by certain gentlemen that this was to be the practice of
the future for all fractures of the patella.   I can only say that, if
ever I am unfortunate enough to break my patella, nothing in the
world will induce me to have my knee-joint laid open so long as I
can get hold of a Thomas's knee-splint and a yard of sticking
plaster.   Such a proposition savours of surgery run mad.   At this
rate, why not at once cut down upon every simple fracture of the
femur, and wire it at once so as to prevent all chance of bad union?

When we have no known means of curing a disease, I am agreeable to any amount of desperate surgery. When we have a known and tolerably certain means, then people's lives should not be put in any jeopardy whatever. Where a limb—by bad treatment or the patient's carelessness—has become absolutely useless, of course Sir Joseph Lister's proceeding is not only justifiable but necessary. But the first point to recognise is, that no useless limbs need ever, under any circumstances, occur, if only patient and surgeon are agreed as to what is to be done, and will carry the treatment out. Eight months without bending the joint is all that is wanted to make a perfect cure in every case. It may be said that a poor man cannot lie up for eight months. But, as has just been pointed out, nobody wants him to do so. At the end of two months let him get up and be fitted with a Thomas's calliper splint, clipping into the sole of his boot, and he can go about his business as well as ever, unless he is in the acrobatic line. It will be far cheaper for the hospital to fit him up with this splint, and give it to him, than to have to take him in six months afterwards for the purpose of having his joint opened and his patella wired.

Compound Fracture of Upper and Lower Maxillæ.—A healthy man of 23 knocked down by an engine; the lower maxilla so smashed that all between the bicuspids of opposite sides had to be removed as loose fragments; the left upper maxilla broken into several pieces, but none requiring removal. Singular to say, the only external injuries to the face was a small cut on the lower lip, and a little damage to the upper prolabium—very much as in the case of Thomas Beckett, mentioned at page 17. Extensive scalp wounds. Delirium and great exhaustion going on for a week, with much putridity in the mouth. Death caused by a sudden hæmorrhage from somewhere about the broken lower jaw. The vessel could not be found at the autopsy, owing to the sloughing condition in which the soft parts were.

Compound Fracture of Humerus into Elbow-joint.—Powerful man of 28, run over by a cart. Humerus smashed in the lower third and split vertically into the joint; large external wound, biceps torn across, and great bruising of soft tissues. The only point that induced one to keep the arm on was the fact that the brachial artery was uninjured. About one and a half inches of the upper fragment sawn off to get bones into good position. Slow but steady recovery, with a stiff elbow, but very useful arm.

Compound Fracture of the Femur in a feeble woman of 46, run over by a cab. Upper fragment protruding through a wound on inner side of the thigh. Reduction and immediate sealing up of the hole with Friar's Balsam. Extensive venous oozing beneath the skin, apparently from injury to the saphenous vein. Treated on Thomas's splint. Excellent recovery, without appreciable shortening.

Compound Fracture of the Femur in middle third, in a youth of 17, from a machinery accident. Profuse suppuration, necessitating extensive openings. Death and removal of a large piece of the shaft in all its entirety. In order to obtain union, a steel drill

Necrosed portion of shaft of femur, from case of compound fracture.

was sent in obliquely through the fragments, going in just outside the femoral artery, and coming out on outer side of thigh. By this they were held in position for nearly two months. After seven months' treatment the patient left with a perfectly useful limb. He is now working at a trade, and only wears a high-heeled boot.

Compound Fracture of Femur—Two Fatal Cases.—The patients died from shock, one within five hours after admission, the other within sixty hours—the fractures in both cases being complicated with other very grave injuries. In one of the cases, among other lesions, there was found the very rare one of a complete dislocation backwards of the carpus without any injury whatever to the radius or ulna. This was proved by *post mortem* dissection. The man's injuries were produced by a fall of earth while excavating a clay-pit. Hamilton says he has only seen one case of this dislocation, and it was produced by a fall on the back of the hand.

Compound Fracture of the Leg-bones—Nine Cases, with One Death.—The fatal case occurred in a man of 60, utterly used up with chronic alcoholism, and in a state of *delirium tremens* when the accident occurred. The fracture was not originally compound, but he made it so by getting out of his bed in his own house on the night of the accident, and walking about till the bones came through. Amputation would have finished him at once, and, after twelve days, he died from delirium and traumatic fever.

To show the severity of some of the cases, the following brief notes of four may be given.

(1.) Thomas Evans, aged 30, was brought in from the country with a compound fracture of both bones in their lower third, produced by a fall of coal from the roof of a mine. The tibia was much comminuted, and the upper end of the shaft stuck out of the wound sharp and irreducible. The fibula was broken in two places. The muscles were greatly lacerated, and the posterior tibial artery torn through. I proposed amputation below the knee: but, as the man refused, the wound was freely enlarged, all loose fragments of tibia were removed, and the protruding portion of shaft sawn off. The most careful antiseptic treatment was employed, and the patient never had the slightest fever. He was five months in hospital, encountering all the usual ups and downs which a severe injury of the sort, attended with the loss of a considerable amount of bone, always necessitates. It was a year before he could work, but, eventually, there resulted as good a leg as ever, with about two inches of shortening.

(2.) William O'Reilly, aged 34, fell on the street, and produced a compound fracture of both bones of the leg in the middle third. The wound was enlarged, and a fragment of tibia removed. Antiseptics employed. All appeared to be doing very well; but, on the 30th day, the dressing being undone for the purpose of renewal, to our dismay there was about half a pint of blood in it, and bright blood was pumping out of the half-healed wound. The man said he had felt a peculiar warm feeling come on about ten minutes previously. Under ether and antiseptics, the wound was freely enlarged, and the fracture exposed. The blood was then found to be welling up between the fractured ends from somewhere behind the shaft of the tibia. To reach the spot seemed almost impossible, and I feared that amputation would be the only resort. However, I literally doubled the leg upon itself at the line of fracture, and so thrust the lower fragment out. About two inches of this I sawed off, and immediately behind was seen an aperture from which the bleeding came. Scratching down upon this, we found the posterior tibial artery with a hole in it. It was divided cleanly through, and the ends tied. Then we found that the hinder end of the fragment just sawn off was excessively sharp, and had actually cut its way back into the artery, and opened it. The parts were carefully replaced, and the man never had a rise of one degree in temperature, nor did he seem to feel that anything out of the way had been done. He left the Infirmary in three and a half months, with his leg in plaster of Paris, and excellent union between the bones. Tying the posterior tibial from the front of the leg by removing a bit of the shaft of the tibia, is not likely to become a popular operation, but it was eminently successful here. If anything at all will prove the value of antiseptics, such a case as this ought to do so.

(3.) William C., aged 36, was struck by a chain cable about four inches below the knee, and both bones broken not far from the joint. The wound was enlarged, and a loose fragment of tibia removed. The tibia was found to be split vertically into the joint. In order to prevent tension, and allow of the fragments setting easily into position, a slice, about half an inch thick, was sawn from the lower fragment. Under antiseptics, the patient recovered without a single bad symptom, and left the hospital with a nearly sound leg in two months.

(4.) F. W., aged 48, was sent from the country, having had both legs smashed by the fall of a heavy piece of metal upon them three days previously. The wounds were stinking, and so were enlarged, and free drainage provided for. All kinds of splints were employed; but, in spite of everything, the bones seemed determined to stick out in all directions. Worn out with pain and suppuration, the unhappy patient was finally—after seven weeks—seized with erysipelas in both legs, and brought to death's door—rigors, delirium, brown tongue, temperature running between 103° and 105°, and pulse from 120 to 130. In despair, all splints and apparatus were cast aside, and the legs were wrapped in lead and opium lotion, and laid between sandbags. He slowly struggled out of his almost desperate condition, and finally recovered. The legs for the rest of the time were simply kept between the sandbags, and allowed to mend in any way they liked. He was seen eighteen months after the accident, and was found to be in full work, with a very useful pair of legs, albeit by no means artistic to look at. I would call attention to two points:—(1.) The advantage of leaving off splints. Very often in the treatment of a long and harassing compound fracture the patient gets worried to death by his splints. Do as you will, they torment him by day and by night. The greatest relief is often gained by leaving them off for a few days, wrapping the limb in a soothing lotion, and steadying it with sandbags. The patient gets a chance of moving himself a little and procuring an easy position to rest in, and so he regains his sleep and his appetite, and, when all right, can bear the splints again with equanimity. (2.) The second point is the wonderful way in which bones will unite without any splints at all, and even when they are constantly moved about. It has led me, as I have already mentioned, in many instances where union seemed to be slow, to take off the splints for a while, and give the ends of the bones some play, so as to set up a little action.

Irreducible Dislocation of the Humerus.—The patient, an elderly woman of 65, put her shoulder out between five and six weeks before admission. A fair attempt at reduction under ether was made but failed. Although a considerable time was spent in manipulation and in twitching to break down adhesions, direct force was only used with very great care. I have once seen the axillary vein torn causing a fatal result; in another case I have seen the artery torn and an aneurism formed; and in another instance the patient died, and at the *post mortem* his ribs were found stove in. Not long ago I myself broke the neck of the humerus in an old lady, although only using my hands.—After a few accidents of that sort one gets cautious. One of the regulation dishes in every medical journal is the " Case of reduction of old standing dislocation," by the constant reading whereof one is tempted to think that every dislocation one meets, however old, ought and must be reduced. This temptation should be sternly resisted, summoning to memory the fact that for one old standing dislocation that is reduced and published there are many which are not reduced and not published, and a certain proportion where serious damage is done, and where the only publication is when the case comes into a law court as one of malpraxis. After all, even in strong men, upon whom abundant force can safely be used, the results of reducing the humerus after it has been out for two or three months are rather disappointing. Stiffened joints almost always. I have succeeded twice in men after intervals of eight to twelve weeks, and in both instances there was hardly any movement. So little do I care about these attempts that when recently a big, powerful man came in with a dislocation of a year's standing, so bad that the arm was absolutely helpless, I made no attempt at reduction, but excised the head of the humerus at once. He is getting a useful and movable joint.

Dislocation of the Metacarpo-Phalangeal Joint of the Forefinger.— The patient, a sailor, about two months before admission had the first phalanx of his right forefinger dislocated backwards on the dorsum of the round head of the metacarpal bone. An immediate attempt and several succeeding ones were made by a medical man to reduce it, but unsuccessfully, and so it was left in the hope that, perhaps, it would turn out a useful finger after all; but it didn't. The phalanx was quite immovable, so that the patient could not shut the finger on the palm. Moreover, the metacarpal head projected so much beneath the skin of the palm that he could not grasp a rope. This finger was, therefore, not only in the way and useless, but actually prevented him getting his living as a sailor. In the case of dislocation of the metacarpo-phalangeal joint of the thumb the difficulty of reduction is often very great indeed, and the causes thereof have given rise to considerable discussion. Much stress has been laid upon the resistance offered by the lateral ligaments and by the powerful short muscles which centre in the sesamoid bones, between which the dislocated metacarpal head is locked as a button is fastened into a buttonhole. But in the case of the forefinger there are no such muscles and ligaments to be overcome, and, moreover, in the two or three cases which I have seen there was no great difficulty in reduction. With steady pulling the phalanx could quite readily be brought on to the end of the metacarpal bone, but the difficulty was to keep it there. The moment the extension was removed it gently slid back to its old place. Whatever may be the cause of the difficulty with the thumb, the obstacle in the case of the fingers is undoubtedly the tearing away of the anterior ligament of the joint, and its subsequent interposition between the cartilaginous surfaces. This ligament, commonly known as the glenoid ligament of Cruveilhier, is a thick fibro-cartilaginous plate lying in front of the joint and having the flexor tendons passing over it. Each plate is attached by its margins to the lateral ligaments of the joint; but the most interesting point about it is that, while it is very firmly fixed to the base of the phalanx, it is very loosely attached to the front of the neck of the metacarpal bone. The result is that, when the forefinger is doubled back on the dorsum of the hand, the ligament is torn from its metacarpal attachment and dragged forcibly over the head of

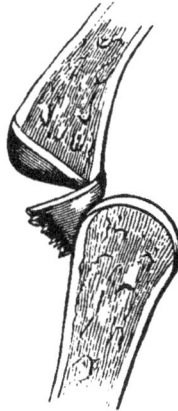

the bone, lodging between the articular surfaces. To induce it to return to its former position is impossible. In two previous cases, one under my own care and the other under that of one of my colleagues, the patients were seen shortly after the accident. The condition being recognized, an incision was made over the palmar surface of the joint under antiseptics, the joint exposed and with the aid of a sharp hook the displaced ligament drawn out from between the cartilages and restored to its proper place. In each case it had not only been pulled off the metacarpal bone, but also split vertically to a certain extent. Movable joints resulted. Fortified by this experience I had no hesitation in cutting down upon the joint in this case. But when an attempt was made to pull out and replace the ligament this was found impossible. It had been considerably split up at the time of the accident and the period which had intervened since then had permitted it to contract such firm adhesions to all the surrounding tissues that it could not be separated as a consistent whole. Under the circumstances it was deemed best to remove the head of the metacarpal bone and so to perform a sort of excision. The flexor tendons being replaced the finger was put upon a splint and very careful antiseptic dressing kept up. The patient left with the parts sound in three weeks. He was seen twenty months after the accident and it was hardly possible to tell that anything had ever been done, the little scar being barely visible. The movements were perfect, and he said the finger was even stronger than its fellow of the opposite side. He was at sea before the mast and could grasp ropes and go up the rigging as well as ever. From this it may fairly be deduced that in the form of dislocation under consideration the proper treatment is to cut down upon the joint antiseptically and try to bring back the cartilage to its normal position:—failing that, to remove the head of the metacarpal bone.

Compound Dislocation of the Elbow—Two Cases.—Entering a ward one day I found a lad of 17, who had just come in, having sustained an injury to his left elbow. On cutting off his jacket sleeve the lower end of his humerus was found sticking out for a couple of inches through a wound at the back of the joint. Before the days of antiseptics excision of the joint would have been the only treatment. As it was, the bone was replaced and complete healing took place under antiseptics without a drop of suppuration. Having a great fear of chronic disease of the joint being set up, if by ill chance an inflammation in its interior occurred, I kept the arm immovable in a splint for a considerable period. After this was removed, the patient was so timid and intolerant of pain, that nothing would induce him to move the joint in the slightest degree. The result was that, although, under ether it was forcibly flexed and extended several times, only very limited movement was ultimately obtained. The lad was seen about a year afterwards. He was working as a light porter, and said that his arm was thoroughly useful, although rather stiff.

In the second case the patient had a distinct dislocation of the bones of the forearm backwards and outwards, as felt by the finger introduced into the wound—not a very common condition. There was sharp bleeding from the interior, and so the wound was enlarged for drainage purposes. Under antiseptics, complete healing occurred without any suppuration whatever, or a rise of a degree in temperature. Taking a lesson from the previous case,

I did not put on any splint. The gauze-dressing sufficed for steadiness, and the joint was carefully moved morning and evening from the very first day. He left hospital in a month. He was seen six months afterwards with the joint absolutely perfect, and a small non-adherent cicatrix about two inches long.

SIMPLE TUMOURS.

1. Giant-celled sarcoma of the lower jaw, originating in an epulis.
2. Fatty tumours—6 cases.
3. Gigantic fibro-cystic tumour of the mamma.
4. Adenoid tumour of the mamma.
5. Adenoid tumour of the cheek.
6. Vascular growth on the tongue.
7. Ovarian tumours—2 cases—(no operation).
8. Fibro-cellular sarcoma from sheath of forearm extensors.

A Tumour of the Lower Jaw, in a healthy, ruddy boy of 15, was of considerable interest. Three years previous to his admission he had pains in the second left lower molar, with some swelling around it. The tooth was drawn, but the swelling did not go away; so he came to the Infirmary, and there was found a very distinct epulis, which I removed very freely with a double-gouge forceps, taking away the full depth of the alveolus and something more. He remained apparently well for five months, when the body of the inferior maxilla—just below the former site of the epulis—began to enlarge, and slowly continued to do so, until a tumour about the size of a pigeon's egg resulted. The outer wall of the bone was greatly thinned, and at the site of the former epulis sprang a low granulation. A probe pushed through this went down into the substance of the growth, which evidently occupied the interval between the outer and inner tables of the jaw, pushing outwards the former only. Now came the serious question, Was this a malignant growth? If so, then a large portion of the lower jaw ought to be removed. Somehow or another I could not make up my mind to this view of the case, and resolved before doing this to try something less serious. Accordingly, under ether, with a strong Volkmann's spoon, I plunged through the granulation tissue at the site of the epulis, and entered a cavity filled with soft material, which I proceeded to scoop out. The bleeding was really alarming; but the more it bled the more vigorous was the scooping, till, in a few minutes, there remained a clean cavity, with smooth bony walls, and not a particle of soft stuff to be felt. It was tightly packed with oiled lint. The soft material being examined microscopically, was found to consist of a tissue not to be distinguished from granulation tissue, with a great many giant marrow cells in it. This was, certainly, not very encouraging, occurring in so young a subject, and one began to think that an error had been committed in not doing the sweeping and more thorough operation. However, by degrees the cavity began to fill up, and its walls to fall in. The boy himself learnt how to pack it with a strip of lint dipped in red lotion, and soon went to his work. Six months afterwards he came to see us, and the walls of the cavity were found to have greatly come together, while the cavity itself was almost filled up. The boy's general health was excellent. He has remained free from any return of the disease.

The treatment and prognosis in this case were certainly very difficult to decide upon. Were the tumour malignant, then anything short of a free removal of a large section of the maxilla could only be a farce. As for the prognosis, even when the microscopic examination was made, it did not clear up matters much. For although the giant-celled sarcoma is probably the least malignant of all the sarcomata, yet the least malignant of them are treacherous. I relied, as I always do, not upon any microscopical examination, but upon clinical features, notably the fact of the tumour having occupied a considerable time in growth, and the fact that it almost certainly arose from some remnants of the former epulis. I remember well, as a student, Professor Hughes Bennett imagining that he had made out a characteristic cancer cell. He strongly urged all surgeons to have a microscope in the operating theatre, and, on the removal of a doubtful tumour, to put some scrapings at once beneath it, and so settle the question of malignancy or the reverse, and act accordingly. I also remember well the profound contempt with which Professor Syme—whose powers of diagnosis were unrivalled—received this proposition; and the more I see of tumours the more confidence do I place in the view which an experienced eye and finger takes of a fresh cut section. Not that I wish to decry the use of the microscope as an aid to diagnosis, but that clinical history and naked eye inspection ought, in my opinion, to take the first places in point of importance

Fatty Tumours.—Concerning the six fatty tumours, the only in-explicable point to me is why such growths should cause pain and weakness in a considerable number of cases. In the case of one female, for instance, the tumour—which was about the size of a lemon—was situated over the last three ribs in a line with the right axilla. She complained of a good deal of pain in the lower part of the right side, "working up" to the back of the right shoulder. It troubled her greatly at night, depriving her of sleep, insomuch that her general health was not a little interfered with. It made both right arm and right leg weak—especially the right arm—so that washing or cleaning in her house was a great trouble to her. After the removal of the tumour all these symptoms dis-appeared. Another female, a cook, who had a lipoma at the upper and back part of the thigh, had to give up her situation on account of the weakness of the limb caused by it. Of course, everyone knows about the weakness and pains caused by super-ficial fatty tumours, but there has never been any satisfactory explanation given, any more than of the weakness of the hand produced by a ganglion. As to actual pressure upon vessels, muscles, or great nerve trunks, there is none. Can it be some reflex action from the involvement of peripheral nerve fibres?

An Adenoid Tumour in the Cheek of a healthy man of 39, was curious from its site. It was first perceived as a little lump about the size of a small pea, beneath the mucous membrane, opposite the second upper molar. It remained of this size for about eight years when it rather suddenly increased to the size of a marble. At this size it remained for about eight or nine years more, and after that it grew steadily till it had attained the size of a hen's egg. It disfigured the patient very much by protruding the cheek so that he was anxious for its removal, which was easily accom-plished. It was found to be an adenoid tumour of exactly the same character as the common adenoid (chronic mammary) tumour so frequent in the breast. It had sprung without doubt from one of the submucous baccal glands, which are identical in structure with the mammary gland.

Fibro-Cystic Tumour of the Breast.—The most important, how-ever, of all the simple tumours was the gigantic fibro-cystic tumour of the breast, of which an illustration is given. The patient was an unmarried woman, about 31 years of age, thin, but not un-

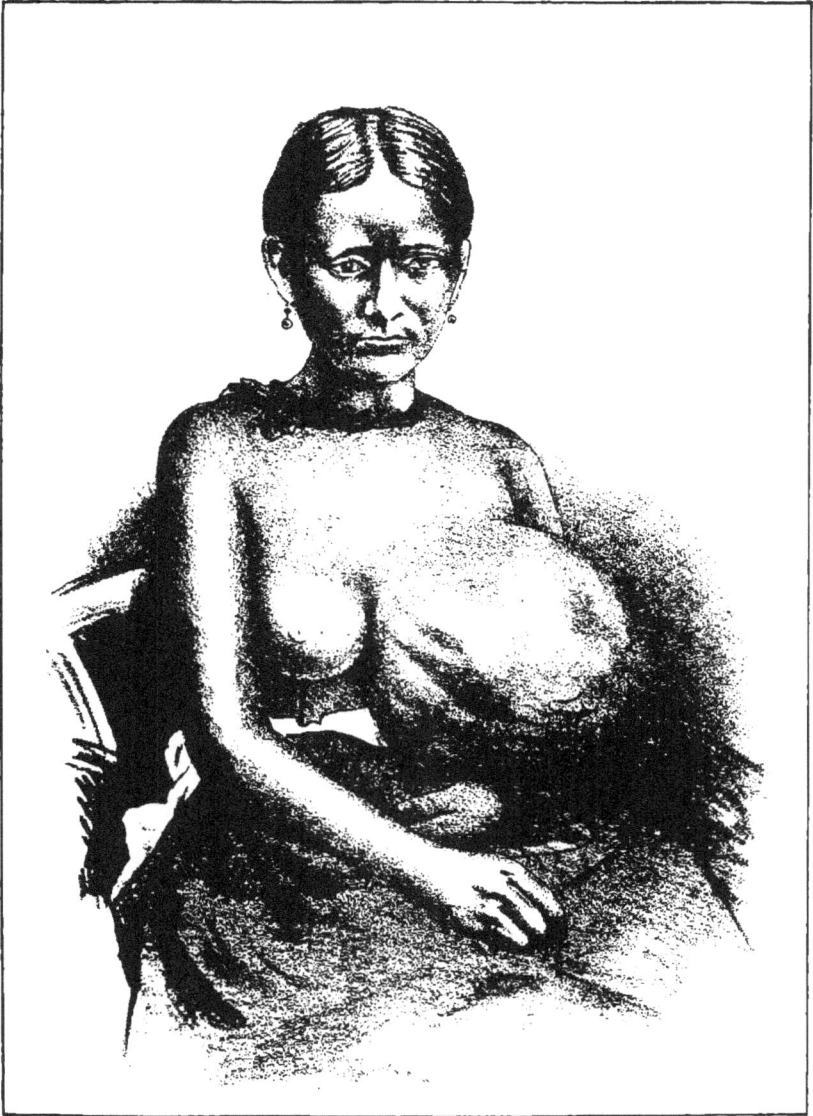

healthy, who lived in a colliery village not far from Liverpool. The tumour took about ten years to attain its enormous size, and at last became unbearable from its magnitude. The unfortunate woman positively had to carry it in her arms like an infant, and could do no work on account of it, seeing that she could never put it down as might have been done with the infant. From its size, general appearance, and feeling, it was easily enough diagnosed as a fibro-cystic growth, and its removal was also readily determined upon. The only dangers to be apprehended were loss of blood and the fear of air entering into certain great veins which ran up from it over the clavicle into the veins at the root of the neck. These were the ordinary superficial veins of the region, but so hypertrophied that some of them, when swollen, were as large as the tip of the little finger. The patient being under ether the tumour was held up by assistants, and an india rubber cord wound tightly round the breast close to the chest wall. This compressed for the moment the great veins alluded to, so that when they were cut across with the first slash of the knife no air got into them and then they were promptly stopped by the pressure of a firmly held sponge. Having made two incisions which should just leave skin enough to cover over the area laid bare by the removal, I simply cut away as rapidly as I could without waiting to take up vessels, sponges being crammed on to the bleeding surface and maintained there. Nevertheless, the hæmorrhage was very sharp for a minute or so, although not so bad as to be seriously alarming. The sponges were then removed one by one and the vessels beneath them tied. The patient made an excellent recovery, without any suppuration, under antiseptic treatment. The drawing of the tumour conveys a better idea of its size and character than any description can give.—Curiously enough she returned not long ago with a very small growth just above the cicatrix of the operation. It was quite movable and obviously was to be accounted for by the fact that a little piece of mammary gland tissue had been left behind, which had developed a growth of the same character as the large one. It was nipped out and found to be an early fibro-cystic tumour. As all breast tissue is now removed no further recurrence need be anticipated.

MALIGNANT DISEASES.

SCIRRHUS OF THE MAMMA.

1	Single, ...	67	Breast and glands,	Remains free 4 years 5 months after operation.
2	Single, ...	50	do.	Remains free 2 yrs. 6 mos. (lost sight of).
3	Married,	47	do.	do. 2 yrs. 6 mos, do.
4	Married,	41	do.	do. 1 yr. 6 mos. do.
5	Married,	41	Breast alone, ...	do. 18 months and died from paralysis.
6	Single, ...	36	Breast and glands,	Recurrence. Death 7 mos. after operation, and 30 mos. fr. commencement.
7	Married,	52	do.	Recurrence. Death 3 mos. after operation, and 6 mos. fr. commencement.
8	Married,	54	do.	Recurrence.
9	Married,	40	do.	Recurrence. Death 10 mos. after operation, and 34 mos. fr. commencement.
10	Married,	36	do.	Died after operation from erysipelas and septicæmia.

Four cases were under treatment, but were too far gone to be amenable to operative treatment.

As the ten cases, which were the subject of operation, form part of a paper on cancer of the breast, contributed to the surgical section of the British Medical Association Meeting of 1882, very little need be said about them. The operations were all sweeping and of great extent—consequently serious. The one death was a source of great mortification, because the patient was practically well and was going to leave the Infirmary on the following day (the twenty-sixth after the operation) when she was seized with a fleeting erysipelas and symptoms of a septicæmic character which carried her off in ten days. Only a small sore, about the size of a shilling, remained to be healed. In nearly every one of the fatal cases, which have occurred in my practice after removal of the breast, erysipelas and septicæmia have been the cause of death.

Attempt to Remove the Cervical Glands.—Patient No. 8 was a thin, spare woman, who remained free from the disease for about fifteen months after operation, but in January, 1882, began to notice the glands above the clavicle enlarging. She, however, never presented herself till the end of April, when she was driven to seek assistance by constant and intense pain in the arm, evidently produced by pressure on the cords of the brachial plexus. One incision was made along the posterior margin of the sterno-mastoid, and another along the clavicle and a triangular flap of skin raised. The external jugular vein was divided, and at a later stage the omo-hyoid muscle. The affected glands were then found lying upon the brachial plexus and subclavian artery, and as many of them were dissected off as possible. They were so intimately adherent to many of the deep structures, however, and to the scalenus anticus muscle, that their complete removal was impossible. A couple of them had grown so completely round one of the cords and enclosed it with so tight a grasp, that the nerve below that spot was swollen to twice its ordinary size, and was red from interrupted circulation. These glands were peeled off the nerve, and from the moment the patient awoke from the chloroform, she experienced relief from her former pain. Had this unfortunate woman returned as soon as she perceived the first slight enlargement the result might have been different from what it was, because the disease simply went on until she died from it a few months afterwards. The difficulty of removal does not seem to be other than can be overcome by any one with a good knowledge of anatomy, and a sufficient amount of patience. If we remove the axillary glands, why not the cervical? The principle is the same.

Since this case, I have removed an enlarged cervical gland, which appeared some months after a removal of the breast and axillary glands. The first gland which becomes implicated lies just at the outer edge of the sterno-mastoid, immediately upon the subclavian vein. I have also cut down upon the highest axillary gland which lies just below the clavicle, by detaching the pectoralis major from the clavicle, and picking the gland off the first part of the axillary vein. When clearing out the axilla a few months previously this gland had been missed, being the highest of all.

Imperfect Operations.—In cases 6 and 7 the operations were known to be incomplete, and therefore a cure was never anticipated. The breasts were cleared away easily enough, but when the highest axillary glands were reached, some of them were so adherent to the vein that they could not be removed without opening it. Both patients were weakly women, and evidently so exhausted by the time the operation reached this stage that it was not deemed wise to pursue it any further. Had it been otherwise, there is no reason why the pectoral muscles should not have been cut through near their attachments, and the portion of vein—to which the glands were adherent—removed. The vein has been cut and tied often without any particular mischief resulting. I have thrice removed about an inch and a half of the internal jugular vein, implicated in cervical cancerous glands, without the patients being in the least degree the worse for it. Indeed, they made more than usually rapid recoveries.

TABLE OF MALIGNANT DISEASES, NOT INCLUDING MAMMARY CANCER.

CASES REMAINING FREE AFTER OPERATION,

No.	Sex.	Age.			
1	Male,	68	Epithelioma of lip (middle two thirds),	Removal,	Remains free after 3 years.
2	Male,	60	Rodent ulcer of the eyelid,	Do.	Remains free after 3½ years.
3	Female,	47	Epitheliomatous ulcer of foot,	Do.	Remains free after 4 years.
4	Male,	56	Epitheliomatous ulcer of neck,	Do. involving the carrying away of a piece of trapezius muscle,	Remains free after 3 years and 4 months.
5	Female,	30	Epithelioma of clitoris and nymphæ,	Do. of the growth, and also of the inguinal and femoral glands of both sides,	Remained free, and died from acute phthisis five months after operation.

CASES IN WHICH THE DISEASE RECURRED.

No.	Sex.	Age.			
6	Male,	52	Epithelioma of the cheek,	Removal of part of cheek, side of tongue, anterior pillar of fauces, tonsil, portion of lower jaw, sub-maxillary, and cervical glands,	Died 9 months from commencement of disease.
7	Male,	44	Do.	of part of cheek and masseter muscle, with alveolus of jaw,	Died 20 months from commencement of disease.
8	Female,	55	Do.	of part of cheek and alveolus. Sub-maxillary glands removed at a second operation,	Died 22 months from commencement of disease.
9	Male,	68	Do.	Primary removal and recurrence. Second removal, involving nearly whole vertical ramus of jaw,	Died 46 months from commencement of disease.
10	Male,	71	of lip,	Extensive removal of lip and clearing out of sub-maxillary glands,	Died 36 months from commencement of disease.
11	Male,	42	of tongue,	Very complete removal of whole tongue. Three subsequent operations on sub-maxillary and cervical glands,	Died 20 months from commencement of disease.
12	Male,	74	do.	Removal of one half of the tongue,	Died 11 months from commencement of disease.
13	Male,	75	do.	do.,	Died 7 months from commencement of disease.
14	Male,	46	of scrotum,	Four operations involving the removal of one testicle, nearly the whole scrotum, part of the skin of the perineum, and the inguinal and femoral glands of both groins,	Died 31 months from commencement of disease.

TABLE OF MALIGNANT DISEASES, NOT INCLUDING MAMMARY CANCER—*Continued.*

CASES IN WHICH THE DISEASE RECURRED—*Continued.*

	Sex	Age	Description	Operation	Result
15	Male,	44	Round-celled sarcoma originating in a gland behind the left ear,	Removal of growth along with part of sterno-mastoid muscle, part of parotid gland, and some cervical lymphatics,	Died 36 months from commencement of disease.
16	Male,	21	Round-celled sarcoma arising at edge of left pectoral muscle,	Removal of tumour and clearing out of lower axillary glands. Second attempt on upper glands through pectoral muscle,	Died 15 months from commencement of disease.
17	Male,	16	Sarcoma of antrum,	Removal of upper jaw, with portions of the temporal and masseter muscles,	Died 7 months from commencement of disease.

DIED AND LOST SIGHT OF.

	Sex	Age	Description	Operation	Result
18	Male,	49	Epithelioma of lip and cheek,	Removal of whole lower lip and part of left cheek,	Recovered—lost sight of.
19	Male,	62	Epithelioma of cheek,	Do. of part of cheek, anterior pillar of fauces, part of soft palate and lower jaw, with sub-maxillary and cervical glands, and a portion of the parotid,	Died of septicæmia.

CASES UPON WHICH NO OPERATION WAS PERFORMED.

	Sex	Age	Description		Result
20	Male,	46	Secondary malignant affection of sub-maxillary cervical glands following epithelioma of lip,		Died 19 months from commencement of disease.
21	Male,	76	Epithelioma of nose, originating in a wart,		Died 15 months from commencement of disease.
22	Male,	77	Sarcoma arising in the skin over the chin,		Died 9 months from commencement of disease.
23	Male,	19	Periosteal sarcoma of base of skull,		Died 8 months from commencement of disease.
24	Male,	58	Sarcoma arising at edge of great pectoral muscle, probably in a gland,		Died 7 months from commencement of disease.
25	Male,	64	Scirrhus of tonsil and cervical glands.		Died 7 months from commencement of disease.
26	Male,	53	Epithelioma of the tongue,		Died 7 months from commencement of disease.
27	Male,	62	Do.,		Died from commencement of disease.
28	Male,	45	Do.,		Died 9 months from commencement of disease.

MALIGNANT DISEASES,

OTHER THAN MAMMARY CASES, UPON NINETEEN OF WHICH OPERATIONS WERE PERFORMED.

Case 1—Epitheliomatous Ulcer of the Foot.—The ulcer was about the size of a shilling, and was situated on the sole of the foot, just in front of the heel. It originated in a warty growth produced originally by a nail in the patient's boot. As usual, it was carefully and persistently irritated with solid nitrate of silver. The poor old man, case 21, had a wart on the side of his nose for some years, which a meddlesome chemist undertook to remove with nitrate of silver—a proceeding which ultimately cost the patient his life, as a rapid ulceration took place, by which nearly the whole of the organ was eaten away ere he applied at the Infirmary for advice. There are two detestable habits which one would like amazingly to see done away with. One is the painting on of tincture of iodine, and the other is the rubbing on of nitrate of silver. No sooner do certain men see a swollen joint, or an inflamed gland, than they set to work to daub it well over with tincture of iodine. This, certainly makes the skin brown and the surface uncomfortable, and in this way may prove a solace to the patient; but beyond this I have never been able to see any effect, although I was brought up to believe in it as a valuable remedy, and, consequently, have come to disbelieve in it by sheer force of experience. As for nitrate of silver, the number of warts and little lip sores which are worried into epitheliomata by it passes belief. In this way it has positively killed more people than it has ever saved by all its other virtues. No sooner does a patient get a sore on the lip or tongue than he immediately has it diligently polished with nitrate of silver till whatever chance it had of coming right of its own accord is speedily destroyed. Or he gets a small abrasion on the penis, and straightway the caustic stick is applied, and a condition of inflammatory induration produced so like a hard chancre that it cannot be told from the real thing. As a stimulant, it is nearly always too strong; as a destructive caustic it is useless. In the vast majority of cases it is a mere useless irritant, and its employment ought to be limited to an occasional wipe over a sluggishly granulating wound.

Case 5—Epithelioma of the Vulva.—The patient, a married

woman, only 30 years of age, had an epitheliomatous growth of
the vulva for about 18 months before admission. It involved the
clitoris and nymphæ, and projected beyond the labia majora, being
about 1½ inches in diameter. It was removed by a sweeping circular incision, which carried everything away down to the bone.
Then the inguinal and femoral glands were removed *on both sides*,
as they were distinctly enlarged. This involved a very extensive
dissection; but the patient made a good recovery, though somewhat slow, and was sent home well to the country in six weeks.
I heard from her medical attendant afterwards that she died
from acute phthisis five months after the operation. Although
she had had a miscarriage in the interval, all the parts were
sound. He was afraid, however, that there was a small femoral
gland enlarged about three inches below Poupart's ligament in
one groin. This would obviously have been so easily removed
that the case has been put in the first category.

Case 6—Epithelioma of the Cheek.—In this patient a great epitheliomatous growth occupied the interior of the left cheek, for the
removal of which the following operation was undertaken. An incision was made from the left angle of the mouth to the sterno-
mastoid muscle, and another in an upward direction along the
anterior border of that muscle nearly to the ear. The lower jaw
was removed from the first bicuspid up to the coronary process, and
with it the submaxillary lymphatic glands, which were much enlarged and adhering firmly to it. The interior of the cheek was
scooped out, and the side of the tongue and the anterior pillar of
the fauces removed. The left tonsil was then dissected away.
Finally an enlarged gland was taken out from beneath the anterior
edge of the sterno-mastoid, which gave a good deal of trouble,
owing to the way in which it stuck to the descendens noni and
internal jugular vein. From this formidable operation the patient
rapidly recovered; but, extensive as it was, it was not sufficient.
Speedily the glands on the opposite side of the neck became
affected, and the disease returned in the cicatrix. The patient died
exhausted five months after the operation.

Cases 7, 8, 9, and 10—Epithelioma of the Cheek.—In all these
cases operations quite as tedious, and hardly less extensive, than
the one just mentioned, were performed. In two of them considerable portions of jaw required to be removed along with the soft

parts. One patient, aged 62, died from pyæmia eighteen days after the operation.

Cases 11, 12, and 13—Cancer of the Tongue.—Case 11 was that of J. W. D., a fine, powerful man, only 42 years of age. He had suffered from syphilis about twenty years previously, and his present complaint commenced about six months previously as a white speck opposite an irritating tooth. When he came to hospital there was a large mass of malignant disease along the side of the tongue. Being anxious to remove the whole tongue thoroughly, I first performed a laryngotomy, and inserted a Durham's canula, afterwards packing the throat with sponges. This only took a few minutes to do. Then I slit the cheek up as far as I could; finally, the tongue was split down the centre, and each half removed as far back as I could go with the knife. On the twelfth day the patient went home. One month subsequently a hard gland was perceived at the angle of the jaw, which was removed. Three months after that there seemed to be one beneath the jaw, nearer the middle line, so a thorough clearance of all submaxillary glands was made. Then ensued a period of seven months, during which there was apparent immunity, at the end of which time he returned with a considerable swelling on the left side of the neck, reaching from the thyroid cartilage to the clavicle. He insisted upon an attempt being made upon this also; but, when the swelling was cut down upon, it was found to be a mere glandular shell, which burst and discharged a grumous fluid. It was utterly impossible to dissect it cleanly away from the great vessels, and so it was plugged with turpentine and left. Of course, the disease extended steadily from this focus, and spread all over the neck, finally exhausting the poor fellow. The tongue and the parts below the jaw remained quite sound to the last. His speech was excellent, and the cheek scar barely perceptible.

In the other two cases (12 and 13) the greater part of the tongue was removed by simply using the gag and splitting it up. In one the disease recurred locally, and in the other in the neck glands. No attempt at a second operation was made, owing to the advanced age of the patients. Indeed, the partial removal of the tongue in both cases was only undertaken on account of the pain of the local sores. The three cases upon which no operation was performed were kept in hospital to try the effect of Chian turpentine, which

E

was then greatly exercising the minds of surgeons. It proved, of course, an arrant failure.

Touching the methods of removing the tongue, I have done the operation by every one of them, except by the thermo-cautery. As might be imagined, various methods suit various cases. Where the floor of the mouth is much involved, Syme's original plan of dividing the lower jaw gives excellent access to the affected parts. When it is necessary to go very far back and make a clean sweep, the plan adopted in J. D.'s case is admirable. The preliminary laryngotomy does away with all danger of choking during the operation, and with the fear of blood being sucked into the larynx, and subsequently setting up an infective pneumonia. The perfect tranquillity of the patient allows the removal to be effected quietly and thoroughly; while the dividing of the cheek gives abundant room to work in. But the splitting of the tongue into two halves, and the subsequent removal of each half separately, is by far the most important advance in the operation which has been made of late years. A valuable wrinkle is the knowledge of the fact that the lingual artery runs through the deep part of the tongue, so that very nearly the entire thickness of each half can be snipped through with scissors from above downwards before the vessel is reached. I have on several occasions cut steadily through till only a small amount of muscular fibre was left, through which I knew the artery ran. This was ligatured with silk thread, and, being cut, no spouting of the artery occurred at all. The tying in of a little muscle along with the vessel is of no great consequence. For minor removals, the use of Smith's gag or the one recently brought out by Krohne and Sesemann, combined with splitting, amply suffices; so that I have given up using the ecraseur as being cumbersome. It is rather amusing to note how Mr. A. employs a straight scalpel to divide the organ, Mr. B. a probe-pointed bistomy, and Mr. C. a pair of curved scissors, and how the intense desire of these gentlemen to march in the van of original discovery induces them to associate their individual names with each of these very uncommon and specially particular methods of operating.

It is worth noticing that only in one of the six cases was a history of syphilis made out. From all one can gather syphilis has no inherent power of originating cancer of the mouth or tongue. If it did so, then cancer of the mouth and tongue would be much more common in women than it is. I have made particular inquiry of late into the subject, and can only find that a limited number of cancerous patients have had undoubted syphilis. Doubtless, these persons had the predisposition to cancerous disease in them already, and only wanted the stimulus of a local chronic irritation to set it ablaze. Such a stimulus is afforded by a chronic syphilitic ulceration or induration of the tongue or cheek, and so in that sense syphilis is provocative of cancer, but only in common with any other irritation.

Case 14—Epithelioma of the Scrotum.—This case deserves a little notice on account of the repeated attacks made upon it at the urgent request of the patient, a most courageous man. Although not really a chimney-sweep, his occupation was much of the same character, being that of a stoker in a gasworks, and he admitted that the wrinkles of his scrotum were generally full of gritty powder. He came with a circular fungating sore, about three inches in diameter, on the scrotum. It was excised, with an apparently wide margin of sound tissue. In doing so, the left tunica vaginalis was opened, as the deep part of the ulcer reached it. Part of it was removed, but there was enough left to enclose the testicle, so it was stitched up with catgut, and no harm resulted. The inguinal glands of the left groin, being obviously enlarged, were removed. Some five months later he came back with an extensive return of the disease. This involved the removal of the left testicle, and nearly the whole of the skin of the scrotum and root of the penis; just enough being left to cover in the remaining testicle. The inguinal glands in the right groin being now affected, were all cleaned out, as well as a group of femoral ones filling up the saphenous opening, which required delicate handling, as they were directly on the coat of the femoral vein. Eight months later he came back with a return in the skin at the very back of the scrotum, just in front of the anus, the ulcer being only about the diameter of a half-crown piece, but very deep. This was swept away, involving a dissection of the bulb of the urethra to get it thoroughly off. After this the scrotum and perineum remained quite sound, but the disease re-appeared well up in the right inguinal canal. At the man's request I made a fourth attack upon this, but in vain. It was impossible to remove it, as it had crept up into the abdomen along the cord. Chimney-sweeps' cancer is noted for its intensely infiltrating power, and certainly the persistence of the disease in this case was horrible. On each occasion it was most fully removed (as our ideas of full removal are at present), only to return. The patient died nineteen months after the first operation, and thirty-one months after the commencement of the disease

SARCOMATOUS GROWTHS.

Case 15—Sarcomatous Growth behind the Ear.—Peter G., a healthy-looking man of 44, about two and a half years before his admission, detected a small, hard, but freely movable lump behind his left ear, between the left mastoid process and the lobule. This is the site of a lymphatic gland, and in this the disease almost certainly originated. It slowly and painlessly increased, until about six months before his admission, when it grew rapidly, and began to cause him considerable distress. On his admission the tumour was of a reddish-blue colour, and about the size of a small tomato, which it resembled not a little in general appearance. The skin over it was extremely thin, almost translucent, and apparently incorporated with the growth, which itself was only slightly movable on the subjacent periosteum. It threw the ear forward, and from its lower part a hard, ridge-like process passed down behind the angle of the jaw. The tumour was removed by a circular incision, part of the ear-lobe being carried away. It was adherent to the periosteum and to the sterno-mastoid muscle, part of which had to be removed. The ridge-like process was cut out also, and along with it a good portion of the subjacent parotid gland, and another piece of sterno-mastoid, as well as some cervical glands from over the sheath of the carotid. Then the mastoid process and remaining area from which the tumour was removed was seared with a thermo-cautery. Nevertheless, I felt that the mischief was not thoroughly cleared away. When once periosteum is affected, bone is affected also, and to remove the mastoid process and a considerable part of the squamous bone was impossible. And so the disease soon sprang up among the granulations which crept over the denuded area, and in a few months attained an enormous size. The unfortunate patient died from pain and exhausting discharge about six months after the operation, and thirty-six from the commencement of the disease.

Case 16—Sarcoma arising near the edge of great pectoral muscle.— This was a very painful case occurring in the person of a fine, healthy-looking lad, aged 21. About eight months before admission he noticed a small, soft, freely movable lump, about the size of a bean, situated one inch outside the left nipple on the edge of the great pectoral muscle. At the end of three months it was as large as an orange, and after that it grew rapidly and the surface gave way and ulcerated. The accompanying drawing shews his appear- ance when he entered the Infirmary. It was clear that removal could only delay the evil day; but the sight and smell of this frightful fungating mass were so horrible that, at the pres- sing request of the patient, I attacked it. It was removed by two elliptical incisions enclosing it between them. It came off without extreme difficulty, leaving some ribs bare and the lower part of the axillary vessels exposed. Some enlarged axillary glands which came into view were removed. Under antiseptic treatment the patient left the Infirmary twenty days after the operation with the whole wound sound except a mere point at the upper end. His condition shortly afterwards is given in the second illustration. In about five months he returned with a swelling beneath the left clavicle. That clearly meant that the highest axillary glands had now taken on the disease. The original cicatrix being perfectly sound and the lad apparently in excellent health, it was impossible to resist the desire to make one more effort for him, although it was felt to be a forlorn hope. A vertical incision about five inches in length was made from the clavicle downwards. Then the great and small pectoral muscles were both cut clean across, and the whole length of the axillary vessels and brachial plexus below the clavicle laid bare, and with them the diseased glands. They formed a soft, friable mass, which broke down and bled furiously at the slightest touch. It quite surrounded the vessels and nerves, and had infiltrated the thoracic wall. As much as possible was scraped away, but of course no proper removal was possible. The pectoral muscles were sewn together with catgut stitches and the wound closed up. In sixteen days the patient went home with everything healed. He died about six weeks afterwards, the disease having spread with terrible rapidity upwards beneath the clavicle into the neck. At the autopsy the scars of both wounds were quite sound. Both lungs were full of malignant deposit, but there was none in the liver. The pericardium was full of fluid that seemed sero-purulent to the naked eye. The left axillary vessels were sur-

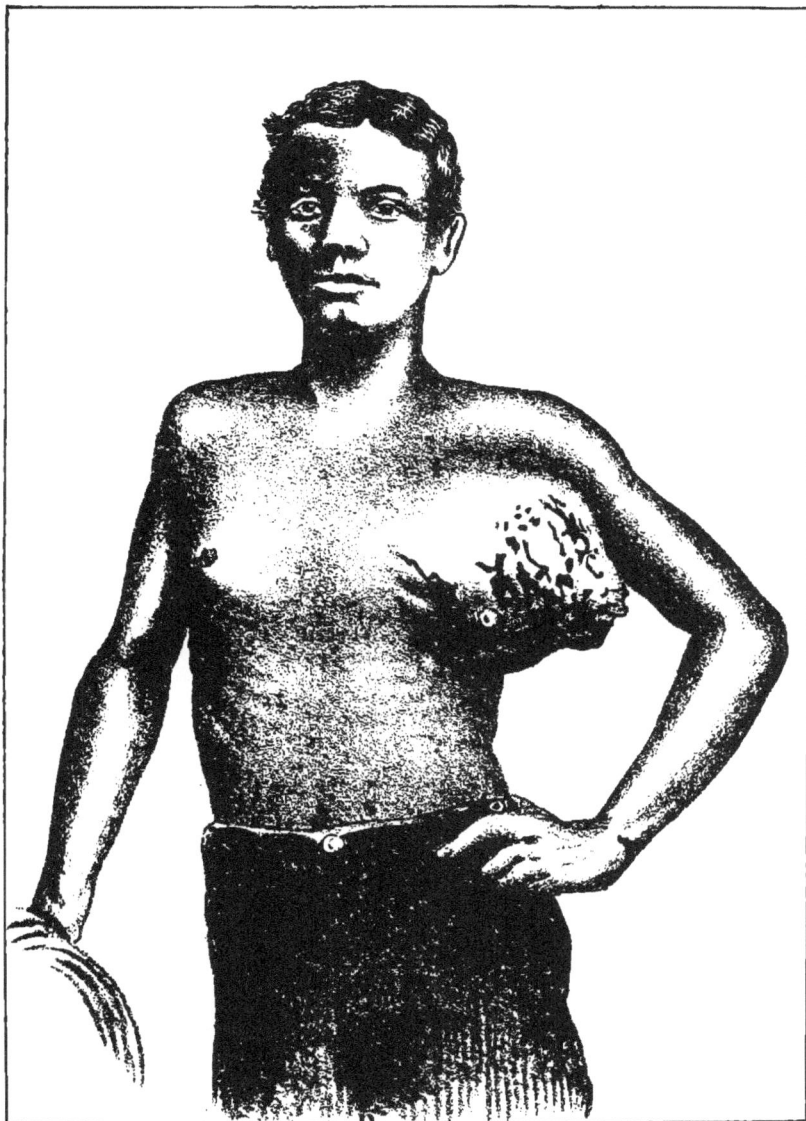

rounded by a mass of disease which extended up into the neck. The internal jugular vein was blocked with soft friable deposit, evidently malignant. So ended this case like all the others. The original tumour was a round-celled sarcoma, and most probably originated in the lowest axillary gland which lies just under the edge of the great pectoral muscle. I have seen primary scirrhus in a female arise in that gland, while the breast never became affected at all.

Case 17—Sarcoma of the Antrum—Removal of the Upper Jaw.— Two months before his admission the patient, a fine, healthy lad of 16, complained very much of toothache in a left upper molar, and in about a fortnight his cheek began to swell. When he arrived this apparently simple and almost painless swelling was diagnosed to be a very rapidly growing sarcoma originating in the antrum. The upper jaw was removed with very little expectation of permanent immunity. No sooner were the soft parts thrown back than the disease was found to be much more extensive than was imagined. The whole of the superior maxillary and malar bones were removed, as well as the zygoma and a good part of the temporal and masseter muscles. The hæmorrhage was very sharp. Notwithstanding a removal of the most extensive and sweeping nature the disease returned in a very few weeks, and he died within five months from the time of the operation.

Case 23—Periosteal Sarcoma of the Base of the Skull.—Cases of this disease are so rare as to make a few details of the present instance not unworthy of note. The patient was a young man of 18, a farm labourer, who, until his disease attacked him was always stout and enjoyed good health. He was admitted on January 24th, 1880. In June, 1879, he began to complain of pain in the head on the right side and of deafness in the right ear. In August, a lump appeared on the left side of his neck, and in October similar lumps came on the right side. They all increased with rapidity. In November, he began to have difficulty in swallowing and speaking and his nostrils became blocked up, During July and August, he suffered a great deal of pain in the head, but not much afterwards; still by reason of inability to swallow he lost flesh greatly and became very weak. When he came to the Infirmary he was pale and thin, and hectic. On the right side of the neck was a mass of enlarged glands as large as a fist, elastic and resistent but not very movable. There were two or three small loose ones beneath the angle of the jaw. On the left side of the neck was a condition exactly similar to that on the right, only the glands were not so large. He could only partially open the mouth, but, when he did so, the soft palate was seen to be bulging, and tense on the right side, while the uvula was pushed over to the left. On passing the forefinger up behind the soft palate it was found to be free, but pushed forwards by a large, soft, fungating mass which projected forwards from the pharyngeal wall. This mass was very friable and bled freely when touched. It completely blocked up the nostrils and produced difficulty in breathing, speaking and swallowing. Indeed, he could swallow nothing but bread and milk or beef tea. He was quite deaf in the right ear, while a profuse mucous discharge was secreted from the nose. Several times a minute the right eye spasmodically closed. On touching the cheeks with a pin it was found that sensation was decidedly defective over the right cheek from the outer extremity of the eyebrow to the angle of the mouth, but there was no loss of muscular power. If the patient wanted to get into the sitting position as he lay in bed, he first rolled on to the left side, and then, fixing his forehead upon the pillow, he pushed himself up with his arms till he became erect. He then turned his head and body in one piece. After a week or two the glands on the right side of the neck softened down, and an incision being made, a considerable amount of very healthy-looking pus came from them, after which the patient for a time experienced decided relief in swallowing and breathing. During the month of February a copious discharge of creamy, "laudable" pus drained from the incision; but some dropping down towards the clavicle, a small lower counter opening had to be made. Soon the left glands also broke down, suppurated, and were opened. Discharge, pain, and general distress wore the patient out, and he died about seven weeks after his admission. At the autopsy the growth behind the palate was found to have originated at the base of the skull, probably in the neighbourhood of the body of the sphenoid. It had spread forward in front of the upper part of the spine, extending into the nares in front, and fungating on the posterior wall of the pharynx. It was pale, watery-looking, and very friable. It did not seem to have laid serious hold upon the bones, but could be

stripped off them, leaving them bare and smooth. It was, therefore, a periosteal sarcoma, and was microscopically demonstrated to
belong to the small, round-celled variety. There were no deposits
in the lungs, but about a pint of thick, grumous, evil-smelling pus
in the right pleura. The liver was studded all over with cancerous
nodules, varying in size from a pea to a walnut, and its interior
was full of them also. In the spleen were a few similar nodules,
and, curiously enough, on the periosteum of one of the ribs was a
solitary specimen. They were all microscopically of the same
character as the primary tumour.

Among the points which attracted our attention was the fact
that although the first symptoms—pain in the side of the head
and in the ear, were on the *right* side, it was the lymphatic glands
of the *left* side which first enlarged. Again, the glands, when they
suppurated and burst did not discharge foul, ichorous, shreddy pus as
one might have expected from glands secondarily affected, but very
creamy and wholesome looking fluid. They almost drained themselves away. In four cases of cancer of the base of the skull,
which I have watched to their termination, in every one the earliest
symptoms of the disease was deep seated pain in the head and deafness in one ear.

Axillary Sarcoma—Another sarcoma occurred in a powerful man
of 58, who, five months before admission accidently discovered a
small, freely movable lump about the size of a pea in the lower part
of the left armpit. This grew steadily, and was soon followed by
enlarged glands in the armpit and above the clavicle. By the time
he came under observation the lump had become a great tumour,
filling up the lower part of the axilla and firmly adherent to the
chest wall, while the upper axillary and the cervical glands were so
enlarged that the clavical could hardly be seen. The left arm was
œdematous, but the amount of œdema varied greatly. The great
cancerous mass rapidly increased and after about a month the
patient was taken with pain in the chest, and "pulmonary
symptoms," and died rather suddenly. After death the nerves of
the brachial plexus were found absolutely lost in the cancerous
mass that filled the axilla. The axillary artery was not much
compressed, but the vein was seriously flattened out. A mass of
cancer, as big as a considerable apple, lay at the root of the left lung
and passed inwards into the substance of the organ, where it

became lost, having no defining edge. .It, therefore, was not a gland secondarily enlarged but a pulmonary growth. The pancreas was one elongated malignant mass, all but its head, which was unaffected. The various growths, which were all of the same cha- racter, consisted of small round cells with linear arrangement such as is seen in young foetal structures:—lympho-sarcoma. It is curious to note how a man may have a great cancerous mass in his lung and have his pancreas transformed into cancer, all but the head, and yet show no signs of such internal mischief. In the observation of many cases of œdema of the arm from the pressure of cancerous glands I have often been struck with the varying amount of the swelling. It varied greatly in the present case. Now if the œdema depended entirely and absolutely upon pressure on the vein and consequent diminution in its calibre, one would have expected it always to keep up to the mark, seeing that the growth did not fluctuate, but kept steadily increasing. The pres- sure on the vein did not notably diminish, why then did the œdema? Furthermore, in operating, the vein has been cut and tied, and no very great œdema of the arm has been noticed. Is it possible that some of the œdema may depend upon pressure on the great nerve trunks? Their influence upon the circulation is admittedly great, and under pressure the currents that pass through them without being absolutely arrested might become variable and so account for a variable œdema. Of course this is a mere specu- lation, but I have never been able thoroughly to reconcile myself to the mere mechanical idea of obstruction of the vein and nothing else.

And now, let us take a general view of the twenty-eight cases mentioned in the table. Twenty-five of them occurred in males, and only three in females. If we exclude the mammary gland and uterus, the remaining parts of the body are admittedly more liable to cancer in the male than in the female. Four cases only occurred in young people, and these at the ages of 16, 19, 21, and 30. In the remaining twenty-four the ages were between 42 and 77.

Between 40 and 50 eight cases occurred.
,,   50 and 60 five   ,,
,,   60 and 70 six   ,,
,,   70 and 80 five   ,,

It may be concluded in a general way that as soon as a person turns 40 he becomes liable to cancer. With regard to the nature of the cases, one was a case of rodent ulcer (which is practically a malignant disease if left to its own devices), one was scirrhus, six were sarcomata, and nineteen were epitheliomata. One was a case of recurrence in the submaxillary glands after a removal of lip epithelioma. Of the patients who have died, those with the sarcomata lived on an average fourteen months, and those with the epitheliomata twenty months from the commencement of the disease.

Out of the twenty-eight cases, nineteen were deemed eligible for operation, and in thirteen instances the operations were of the most extensive character, involving the removal of large masses of tissue, and productive of severe hæmorrhage. Only one patient died, a man of 62, and that from septicæmia. One has been lost sight of. What of the other seventeen? But four are alive. A fifth died five months after operation, practically free. In the case of the four survivors the disease was limited, and the operations, although not by any means trifling, were unimportant compared with the others. Not one of the cases which demanded extensive removal survived, although no pains were spared, nor risk to the patient shirked, to make that removal as free as it possibly could be made. Indeed, in not a few instances I was myself almost shocked at the hideous spectacles which remained after some of the operations, until the parts were stitched up or covered over. And yet no good came of any of them. It is, indeed, a most dismal and disheartening list. The only consolation is that the cases were very bad ones, in the great majority of which the disease had been allowed to progress for a very long time before assistance was sought. Many of the patients came from country districts, and were ignorant persons, who stuck to their work in spite of the most shocking conditions, until fairly compelled to succumb under pain. The epitheliomata, when caught soon, are, of course, very hopeful as regards a permanent clearing away of the local disease, but they undoubtedly affect the glands at a very early period. So that I feel very strongly that to do any real good it will become necessary to remove invariably the glands beneath the angles of the jaw as a necessary part of the removal of an epithelioma of the lower lip, *whether they are felt to be enlarged or not.* In fact, to apply the same principle to cancerous glands everywhere, which I have urged with regard to

the axillary glands in mammary scirrhus. I fear that by the time
the local disease is extensive and the glands fully involved operative interference is too late, although it may *seem* as if it were quite
practicable to carry all diseased structures away. Our proper plan
will be to catch the patient soon, while his local disease is capable
of certain removal, and to carry away the glands which, although
diseased, are yet too small to be felt. Although the extensive operations which have been described have not been of much service to
the unhappy patients, either in my own practice or, as far as I have
been able to see, in the practice of others, still they have been of a
certain use in demonstrating the possibility of performing removals
of an extensive and prolonged character with safety to life. That
being demonstrated, one need have the less hesitation in performing
a considerable operation for the removal of a small affair. When
we carry away a small lip or tongue epithelioma, knowing as we do
the extraordinary risk of secondary lymphatic affection, is it not
our duty to perform a severe operation, even when the local mischief is slight, and carry away submaxillary or cervical glands,
whether we can feel them or not? Would the increased safety
from recurrence not counterbalance the increased danger of the
operation? Such a proposition seems almost to savour of cruelty,
but I am convinced that it will be the practice of the future. If
the patients can stand all this sort of thing when the disease has
attained a great extent, why should they not stand it with equal or
greater safety when the disease is limited? Without doubt, the
two points are, first to get the patients to apply for relief early, and
secondly, to remove about three times as much as seems necessary.
As for extensive removals for extensive diseases, I find, on looking
over my hospital records for eight years, that I have only succeeded
in saving one solitary patient. He was an old man who, in the
spring of 1874, had a small epithelioma removed from the lower
lip. It never recurred; but, in the summer of the same year, he
came under my care in the Infirmary with a great mass of ulcerated,
fungating glands under the right angle of the jaw. To remove it, a
large piece of sterno-mastoid had to be carried away, together with
the digastric and stylo-hyoid muscles. The facial and lingual
arteries were cut and tied, and the descendens noni divided. The
mass was dissected for a long way from the very sheath of the
carotid vessels, and the whole operation was so tedious and bloody

that I more than once heartily repented having begun it. Nevertheless, it was finished, and a great hole about the size of the palm of the hand left open, because there was no skin wherewithal to cover it. It all healed over, and the old fellow is well and at work, nine years afterwards. But he is the solitary survivor of innumerable similar cases. These remarks apply to epitheliomata alone. As to distinctly malignant sarcomata, I regard them as practically hopeless. Only once have I seen a case where a malignant sarcoma did not prove fatal, whether removed or not. The periosteal sarcoma seems a much more malignant disease than the central one arising in the interior of bone.

DISEASES OF BLOOD-VESSELS.

1. Aneurism of first part of aortic arch—proposed ligature of carotid and subclavian.
2. Aneurism of right subclavian—attempted ligature of first part of axillary.
3. Wound of the palmar arch—ligature of brachial.
4. Occlusion of the brachial artery from injury.
5. Phlebitis in veins of lower extremity—three cases.

Case 1—Aortic Aneurism.—The patient, F. J., aged 51, had been for twenty-three years in the artillery, and, for ten years after his discharge, had been night watching and cab driving. Two months before his admission he began to suffer from pains in the right shoulder, in the scapula, and in the back, which he ascribed to rheumatism. They soon became so severe as to compel him to drop work, and he rapidly lost flesh. When he came into hospital there was found at the upper part of the sternum and to the right side a pulsating tumour, which was bounded by the right clavicle above and by the third rib below. The skin over it was reddened. By degrees this extended into the root of the neck above the clavicle, pushing upwards beneath the sternal end of the sterno-mastoid till it came to lie for a couple of inches in front of the trachea. It was diagnosed as an aneurism of the first part of the aortic arch, possibly involving the innominate and tending to increase in an upward direction. The patient's lungs and heart were sound. Before contemplating any surgical procedure a prolonged trial was given to the rest and low diet treatment. The patient was kept absolutely flat upon his back, was fed on milk and beef tea, and had twenty grains of iodide of potassium thrice daily. At the end of several weeks it was found that no benefit had resulted and that the tumour was increasing, while his pains about the shoulder were unbearable. My experience with galvano-puncture has hitherto been so entirely unsuccessful that I saw no good in attempting it. The only project that seemed to afford a prospect of a cure was ligature of the right subclavian and common carotid on the distal side of the aneurism. The danger of this having been fully explained to the patient, he expressed himself willing to submit to any risk provided there was a chance of his getting better. On the day of operation he rose from his bed and walked to the theatre, being the first occasion for a long period on which he had made the slightest exertion. This was much to be regretted. Ether was administered, but it caused a great deal of struggling, insomuch that it seemed impossible to get him sufficiently quiet to commence the operation. He became very livid also, and his respiration became embarrassed by the quantity of froth which continually welled up into his mouth. Chloroform was therefore substituted. This acted better and he was soon quiet, but his breathing seemed laboured all the way through and he gave constant anxiety to those around. The carbolic spray was turned on the neck and the common carotid was cut down upon. There was no difficulty about reaching the vessel and the ligature was passed around it; but I felt so uncomfortable about the patient's breathing that I resolved not to tie the knot, but to wait and see how he got on. He had not had any anæsthetic for some time, yet nevertheless his respirations continued to get slower and slower, while he became more and more livid. We tried to arouse him by all sorts of means, but could not get him conscious. Artificial respiration was persistently worked at and his trachea was opened and a tube put in, so that we might be sure there was no mechanical obstruction. In spite of everything, he quietly breathed more and more slowly, and so to speak, slipped through our fingers in spite of the most vigorous efforts to keep him alive. For the explanation of so painful a catastrophe I was utterly at a loss, but suggested to the audience that under the excitement of the expected operation, together with the exertion of walking to the theatre, the aneurism had given way and the effused blood had pressed on the respiratory organs. It was a limping explanation at the best. At the *post mortem* we found the lungs very emphysematous and the bronchi charged with frothy mucus. The heart was, on the whole, healthy—the left side empty, the right full of dark coloured fluid blood. The aneurism sprang from the first part of the aortic arch. It had pushed forwards so as to erode the sternum, inwards so as to press upon the trachea, and upwards into the neck so as to overlap and press upon the innominate artery. There was no rupture and the sac contained laminated clot. It was not till we turned to the great veins that we found an explanation of the cause of death. The right innominate vein was found quite flattened by the aneurism so that no blood could get through it; the left internal jugular vein was found corked by a firm plug which reached from the base of the skull to its junction with the subclavian and so rendered it absolutely impermeable. This clot by the look of it seemed to be some weeks old. *Thus the two main channels by which the blood is returned from the brain were completely occluded.* So long as the patient lay absolutely motionless on his back the blood from the brain got a sufficient exit by the minor channels of return. But, when additional blood was poured in these were not sufficient. Thus it was that the excitement of the anticipated operation, the rising suddenly out of bed and the walking to the theatre, doubtless threw into his brain a quantity of blood which it could not get quit of. In this condition came the anæsthetic which assisted in this, and so, between the two, his nerve centres became full of unærated blood and ceased to act. From the moment he lay down his breathing attracted attention, so that during the whole operation he was most carefully watched, not merely by the chloroformist, but by one or two of my colleagues. There was, therefore, no mismanagement of the anæsthetic. The mistake was in beginning any operation at all upon a man in whom the cerebral circulation must have been in such a singular state. But as we could not possibly know this, the fatality had to be accepted as one of the misfortunes of war. It certainly would not operate at all in preventing me tying the carotid or subclavian if a favourable case presented itself to-morrow, as such a condition must be so rare as not to enter into one's calculations of risk.* What caused the plugging of the left internal jugular vein I cannot tell. The only satisfactory point in the case was the fact that I did not tie the knot and shut up the carotid. I might then have blamed myself for interfering with the unfortunate patient's already sufficiently embarrassed cerebral circulation.

Case 2, Aneurism of the right Subclavian Artery.—The patient William J., a ship carpenter, aged 40, contracted syphilis twenty

---

* I performed the operation of ligaturing the subclavian and carotid for innominate aneurism a few months ago, with benefit to the aneurism, but not to the patient, who ultimately died.

years previously and suffered at intervals from it ever after. Four months before admission he felt his right arm numb and painful, and in a few weeks a swelling appeared low down on the right side of the neck and simultaneously the right side of the face became œdematous. It is not necessary to detail the patient's various symptoms, but merely to state that, when he came to us we found an aneurism which reached from the anterior edge of the sternum backwards to the clavicle in the posterior triangle of the neck. It pressed so far downwards internally that it was impossible to reach the innominate. It reached so far back externally that the finger could not rest between its outer limit and the clavicle. It was therefore impossible to get at the third part of the subclavian with a view to distal ligature. But it seemed to me that if this could not be got at, why not try the first part of the axillary half an inch lower down? A great objection seemed to be that all the surgical anatomy books describe this operation as a very hazardous and formidable proceeding. To this I paid but little attention, knowing the persistency with which statements about things that are seldom done are copied from book to book. A much more serious objection lay in the fact that there were grounds for fearing lest the artery should be much dilated. I resolved, however, to see what it was like, and so made an incision parallel with the clavicle, divided the pectoralis major and costo-coracoid membrane, tied some twigs of the acromial thoracic artery, and soon laid bare the main vessel and its vein. To my great annoyance, I found it to be very much dilated, and with its coats so thinned and apparently diseased, that I did not venture to tie it. I passed the aneurism needle behind it, and so examined it thoroughly, and convinced myself that it would have given way before a ligature tied sufficiently tightly to occlude it. As there was no possibility of tying the vessel higher up, the consequence of a rupture would have been immediate death from hæmorrhage. Unwilling to risk so dreadful a catastrophe for a benefit that after all was problematical, I closed the wound up. It healed at once, and the patient was none the worse for what was done; indeed, he was somewhat improved, as he confidently looked forward to some ulterior benefit from the operation. He lived for about a month afterwards, and died by reason of the aneurism becoming diffuse. After death it was found to have taken its origin in the first part

F

of the subclavian. The chief points of interest were—(1) The difficulty of making out the condition of the artery behind a powerful, great pectoral muscle. It was suspected to be dilated, but not to have arrived at the distended and diseased state in which it was. (2) The fact that even in a well-built, muscular man there need be neither difficulty nor danger in tying the first part of the axillary artery, if only a free enough incision be made through the great pectoral.

Case 3—Ligature of the Brachial for Wound of the Palmar Arch.—The patient, a youth of 18, fell on a piece of pot, which pierced the palm of his left hand and caused profuse bleeding. The wound was dressed by a chemist, and apparently healed, so that seven days after the accident he was drawing on his glove, while preparing for a drive, when the freshly healed wound burst open, and sharp hæmorrhage ensued. For eleven days various means were employed to stop repeated attacks of bleeding, which occurred as soon as pressure and bandages were removed. They were unavailing. On his admission, ether was administered and the bandages taken off. At once blood welled from the palm, the tissues of which had been during the previous eleven days torn up and disintegrated by blood clots, till now it looked just like a mass of rotten pulp. Without more ado I at once tied the brachial artery under antiseptics, and then scraped and cleaned out the palm, finally swabbing it well with turpentine. The ligature wound was dressed twice, and on the tenth day was absolutely healed without a single drop of pus having formed. The hand was kept elevated on pillows, and treated by the open method, being much wiped and kept clean by absorbent wadding. It granulated all up by degrees, and, when seen some months afterwards, was nearly as good as ever. The wonderfully cleansing effect of the turpentine was most apparent in this instance. I should always tie the brachial at once in preference to tying radial and ulnar for bleeding from the palm. In the present instance the treatment was most satisfactory, and produced the desired effect immediately and permanently. It was very surprising that the tendons acted afterwards, as it was fully expected that they would be all matted up in the cicatrix.

Case 4—Occlusion of the Left Brachial Artery from Injury.—John Gorr, a labourer, aged 49, while shunting waggons, was jammed between the buffers. His left clavicle was broken, and his chest on the same side badly crushed, three or four of the lower true ribs being fractured close to the cartilages. His left upper arm was bruised and slightly swollen, and at the lower part of the belly of the biceps there was a distinct gap, in which a couple of fingers could be laid. Here the muscle had evidently been torn across, and its ends had retracted. The most noteworthy feature in the case, however, was that no pulsation could be felt in the left radial, ulnar, or brachial arteries, although the axillary artery was still beating. The left forearm and hand were distinctly colder, and less sensitive than the right. For about a week the patient's life was in great jeopardy, first from shock, and afterwards from pneumonia brought on by the chest injuries. At the end of that time the axillary pulsation ceased, as if a clot had extended up into the vessel, but in a few more days it returned, and not only in it, but also in the first inch or so of the brachial. We also traced with interest the development of a large vessel, apparently the superior profunda, which became enlarged in order to carry on the collateral circulation. About the same time a very faint pulsation could with difficulty be detected in the radial, as if blood were getting back into it, while the hand became once more possessed of natural warmth and sensation. With rest and good nursing the patient recovered, and, at the end of a couple of months, was discharged fit for work. His condition was as follows :—The interval between the torn ends of the biceps had filled up, leaving an induration to mark its site; the axillary and first part of the brachial pulsated, and so did the radial, while the temporarily enlarged superior profunda had much diminished in size. But the greater part of the brachial artery was evidently permanently occluded. It was pretty clear that the same injury which ruptured the biceps also crushed the artery, the internal coat of which was probably torn across. As a result the vessel became permanently occluded in the same way as the subclavian artery in the case detailed at p. 85. The patient could not be traced after leaving hospital, but it is not likely that his condition would alter much as regards the artery. Fortunately, the utility of the arm was not impaired.

DISEASES AND INJURIES OF THE NERVOUS SYSTEM.

Rupture of the brachial plexus; examination of the seat of injury.
Division of the median nerve; resection and cure.
>   Do.   resection—failure.
>   Do.   no operation.
Stretching of sciatic nerve for locomotor ataxy; marked improvement.
>   Do.   do.   ; no improvement.
>   Do.   for obscure spinal disease.
Stretching of popliteal nerves for leg paralysis after injury.

Rupture of the Brachial Plexus—Exploratory Operation.—William Thomas, aged 30, a coloured seaman, in the beginning of December 1880, fell down a hatchway at St. Vincent. He could not give us any information about his injury, except that he was taken to a hospital, where he was cupped on the chest and bled at the bend of the right arm. He came into the Infirmary on Feb. 5th, 1881. The right forearm and hand were swollen and œdematous. There was absolute loss of sensation in the hand and back of the forearm, and partial loss over the front of the forearm and in the upper arm. There was entire loss of motion from the shoulder downwards, with wasting of the muscles of the shoulder and upper arm. Briefly, then, the arm was completely paralyzed. As the patient stood before us, it dangled helplessly by his side—down to the elbow shrivelled and attenuated—from the elbow to the fingers swollen and doughy to the feel. The faradic current caused the deltoid to move, and also the great pectoral slightly, but no other arm muscle. To account for this paralysis there was no dislocation nor apparent injury of any kind. In the course of our examination it was found that there was no pulse in the arteries of the forearm, nor in the brachial or axillary arteries, the latter of which could be traced as a hard cord right to the top of the axilla, surrounded by the subclavicular cords of the brachial plexus.

Furthermore, there was no pulse in the third part of the subclavian artery, while just at the outer edge of the scalenus anticus muscle was a hard lump about the size of a walnut. It was almost bony to the touch, and had very much the feeling of springing from the upper surface of the first rib. It puzzled us extremely to make out what it could be. The symptoms just detailed left little doubt that some injury had occurred to the brachial plexus above the clavicle, and, as attention has been drawn in such cases to signs indicative of lesion of the sympathetic filaments which accompany it, these were sought for. It was found that on the affected side the right palpebral fissure was narrower than on the opposite, so that the eye looked smaller. This is believed to be due to a paralysis of the detrusor fibres in the orbit allowing the uncontrolled action of the recti to pull the globe backwards. The right pupil, moreover, when exposed suddenly to bright light, contracted quite readily; but when put in the shade did not dilate as much as its fellow, but remained more or less contracted. After a fair time had been allowed to try the effect of electricity, and no good resulted, it was determined to explore the root of the neck in order to find out what had occurred to the brachial plexus, with a view of remedying the injury if possible, and also to discover the real nature of the hard lump just external to the scalenus anticus. Accordingly, on March 2nd, an incision was made along the clavicle, joined by one along the outer margin of the sterno-mastoid. A careful dissection was then made of the subclavian triangle, a work requiring some little delicacy. The hard lump was found to indicate the site where a rupture of the internal and middle coats of the subclavian artery had probably occurred, leading to plugging by coagulation, and subsequent obliteration of the calibre of the vessel. As nothing could be done with this, a search was made for the brachial plexus, and after some time it became clear that the whole plexus had been torn off close to the transverse processes of the vertebræ (which the dissection clearly displayed), and had been dragged down subsequently by the movements of the arm behind the clavicle. Of course, any attempt at re-uniting them was utterly impossible. One fine, single strand of the plexus alone remained untorn, and, when pinched, the deltoid and pectoralis major contracted—the two muscles which, alone of all the arm and shoulder groups, had responded to the battery.

Finding that nothing could be done, the wound was closed and antiseptic dressings carefully maintained, with the result that on the ninth day the extensive dissection wound was absolutely healed without a single drop of suppuration, and with a rise of temperature barely amounting to one degree. The arm remained *in statu quo,* and after some months the patient returned to his native place on the coast of Africa.

The points of interest to be noticed in the above case seem to be:—

(1) That a fall, unaccompanied by any fracture or dislocation, may rupture the subclavian artery and actually tear the strands of the brachial plexus away from the spinal cord.

(2) That the sympathetic nerve symptoms may prove useful in confirming the diagnosis of a torn brachial plexus.

(3) That any attempt to re-unite a ruptured plexus, even when the upper portions of the nerves are to be found, will be extremely difficult, if not impossible, from the fact that the lower portions tend to retreat downwards behind the clavicle to such a distance that the ends, however carefully refreshed and brought together, will not unite.

(4) That if any doubt as to diagnosis remains in the mind of the patient or the surgeon, an exploratory operation is a justifiable proceeding, and, at all events, relieves the mind of the former from that state of uncertainty and anxiety which everyone knows to be more unbearable than knowing the worst.

Three Cases of Injury to the Median Nerve.

(1) *Division of the Nerve—Successful Resection and Union.*— The patient, a blind man, aged 30, earned his living by making mats. He fell through a window, and inflicted a deep transverse cut about half an inch above the left wrist. The wound healed quickly, but it was soon found that the parts supplied by the median nerve were paralyzed. This was six months before he came to the Infirmary. He applied on account of the almost total uselessness of the hand. There was great loss of sensation over the median area, and the muscles of the thumb were so paralyzed that he could not oppose it to the other fingers at all. With regard to the other fingers, the little one had power, but the other three remained in a semi-flexed condition, and quite

powerless. He had to open and shut them with his other hand, and wherever they were put there they remained. Under careful antiseptics, the transverse cicatrix was first laid open, and then a vertical cut was made over the line of the median nerve: a sort of crucial incision, in fact. It was found that the median had been divided, all but one extremely narrow strand on the inner side. This white filament could be made out quite clearly, and also a band of connective tissue which ran alongside it. The ends of the nerve were rounded off in a conical manner, and were about an inch apart. With a very sharp knife they were cut square off, the band of fibrous tissue dissected away, and the refreshed surfaces carefully brought together with three catgut sutures. The uninjured strand, which had been much stretched, was left intact, and it doubled up on itself when the rest of the nerve was brought together. The hand was kept well flexed on the forearm to take off all tension, and the wound was quite healed on the thirteenth day without any suppuration whatever. The hand very rapidly improved, the sensation quickly returning to a great extent, and motion becoming nearly perfect. The patient was seen in March, 1883. The feeling, as tested by a pin, is quite acute over the whole of the palm and fingers. If he tickles the tip of the thumb or of the two and a half fingers supplied by the median nerve, then he has a sensation of "priukling" in the other tips and in the palm of the hand. The muscles have excellent power, so that he can grip nearly as hard as ever. Nevertheless, the thumb and two first fingers have a thinned and atrophied look, and although the sensation, as roughly tested, seems all right, he himself does not find it acute enough to do weaving or mat matting, nor to read raised letters. For a man with his sight the hand would have been as useful as ever, but for a blind man, requiring very special delicacy of touch, it was hardly sensitive enough to enable him to work, although possessing abundant strength.

(2) *Division of the Median—Unsuccessful Attempt at Resection.*— The patient, a school-boy, aged 9, fell upon a glass bottle, and received a deep transverse wound just above the flexure of the left wrist. The wound was treated at first at a dispensary. On admission it was gaping widely, and in a very unwholesome state. The sensation of the thumb and first three digits on their palmar sur-

faces was quite lost, and movement of the thumb was impossible. As it seemed certain that the median was divided, a vertical incision was made over it, across the line of the wound, and the ends were found, which were about an inch apart. I refreshed them, and united them with very fine silver wire, doing the same to one or two of the flexor tendons, which had been divided. The wound was thoroughly cleansed and disinfected, as far as was possible, and treated antiseptically. Most unfortunately, this did not succeed. The tissues were all in an acutely inflamed state, and by the third day the stitches had cut their way through the softened skin, the wound was gaping widely, and the ends of the nerve had separated from each other, owing to the giving way of the stitches. The antiseptics were abandoned, and the wound poulticed. By degrees it slowly healed, leaving a firm cicatrix with the parts all matted to it. The boy was seen about eighteen months afterwards. The hand had regained more usefulness than could have been expected. Sensation over the median area was largely restored, and there was even some muscular power. Movement would have been much better but for the fact that several of the divided flexor tendons had become incorporated with the cicatrix in front of the wrist. The hand was, no doubt, seriously crippled; but possessed infinitely more sensation and power than when he left hospital.

(3) William S., aged 26, was carrying a large sheet of plate glass when he fell, and a large fragment of glass inflicted a severe wound on the front of the left wrist. He came to the Infirmary nine days after the accident, the wound being then granulating. A good deal of suppuration took place about the palm and one or two of the fingers, but by properly planted incisions this was relieved. What nerves he exactly injured one cannot say, but as the wound ran from the centre of the wrist inwards, the median and ulnar were those most exposed. They were undoubtedly both injured, although probably not quite cut through. Sensation on the palmar aspect of all the fingers, but not over the thumb, was greatly diminished; the prick of a pin felt like a "bump." Being warned by the last case, I did not attempt any operative proceeding, but concluded to wait till everything was sound. Nine months after the accident the hand was found to have a lean and wasted look as a whole, the thenar and hypo-thenar muscles being much atrophied. The little finger was stiff and nearly closed. The other fingers were kept in

a half-closed condition. He could not open them with any force, but could shut them firmly. In spite of these defects, however, the hand was quite useful for him in his occupation. I was much interested to find a remarkable return of sensation—all the median area being nearly as sensitive as in the uninjured hand, and the ulnar area perfect. Under these circumstances no operation was required.

At the time when a wound dividing the median nerve is inflicted there can of course be no doubt as to the propriety of suturing the divided ends of the nerve. But any attempt made during the healing of the wound is pretty sure to be unsuccessful, as the ends of the nerve are softened, and will not hold the stitches. The wisest course, then, is to wait until the wound is completely healed, by which time the ends of the nerve will be in a condition fit to hold the sutures. Moreover, the extent to which both sensation and motion (and particularly the former) return, even when there is no possibility of the nerve ends having united, is something remarkable. So that, as in the third case just narrated, the hand, though defective, may be sufficient for all its owner requires, in which case operating is rather a work of supererogation than of necessity. How this sensation and motion return is very inexplicable, except by supposing that it does so through the various junctions which the median has with the ulnar. As some little corroboration of this, I have noticed that, after division of the median, the loss of sensation in the outer side of the ring finger is never so great as in the others. In a paralysis of the ulnar, on the other hand, the anæsthesia of the inner side of the ring finger is not so complete as that of the little finger. This is probably due to the junction of median and ulnar branches at the tip of that finger. According to the results of some dissections of the nerves and muscles of the forearm recently made by M. Verchere, in eleven instances out of fifteen an anastomosis between the median and ulnar nerves was found in the upper part of the forearm. This emanates from the median near the origin of the branches which go to the flexor muscles and joins the ulnar nerves directly, or by the intervention of a small plexus. It results from M. Verchere's researches that the two common flexors of the fingers are supplied both by the median and by the ulnar nerves; and that, therefore, these muscles are not completely paralyzed by division of the median, as has hitherto been considered the case. The anastomosis is admitted, however, to be often very rudimentary.

Two Cases of Sciatic Nerve-stretching for Locomotor Ataxy.—Both patients were males, under the care of my colleague Dr. Davidson, in his medical ward. He requested me to stretch their sciatic nerves. They were both joiners, one (Scott) aged 36, the other (Hume) aged 34. They both had suffered from syphilis, one sixteen and the other fourteen years previously. Scott's symptoms dated from about two years previous to admission, and Hume's from about three years. They were very marked and character- istic, but Hume was decidedly the worse of the two. In order fairly to test the efficacy of nerve-stretching, no treatment what- ever except the operation was employed in either case. Under ether and antiseptics the nerves were laid bare. A hook was put under them which was attached to a spring-weighing machine. which was suspended over the patient. The nerves were then steadily pulled upon until the machine registered 40 lbs., and they were kept at that tension for a few minutes and then released. In both instances the wounds were sluggish and healed very slowly. The patient Hume was much relieved of his pains for a while, but ultimately all his symptoms returned and he died under a year from the time of operation. But the improvement in Scott has been something remarkable and has gone on steadily increasing, although it is now exactly three years from the date of the oper- ation. At that time he could hardly stand, now he can walk four or five miles without fatigue—he is on his legs all day attending to a small shop. He was so blind that he could not see to read and so deaf that he could not hear his watch tick—now his sight and hearing are normal. With his eyes shut he could not stand unsupported, nor could he touch his nose with his fingers—now he can walk eight or nine steps with closed eyes, and touch his nose quite accurately. Then he had severe gastric crises and light- ning pains—now he has only occasional attacks, of what he calls "biliousness and indigestion," followed by some pain, not excessive, in the legs. He has had no physic nor treatment whatever except the nerve-stretching.

Paralysis of the Leg Muscles from Injury—Stretching of the Pop- liteal Nerves.—The patient was a big, stupid looking collier, aged

22. He ascribed his condition to an accident which he sustained a few months previous to his admission, when an iron boiler plate, of great weight, fell upon his right leg. As there did not happen to be anybody near him at the moment, it lay upon him for about twenty minutes, when three men came up, whose united exertions were required to liberate him. No wound was made, nor bones broken, but the limb remained useless from the knee downwards. The affected leg was colder than the sound one and of a blueish colour. There was a little pitting on pressure over the dorsum of the foot. Round the thickest part of the calf the measurement was one inch less round the right than round the left calf: muscular power below the knee was quite gone, so that he could not even move the toes.—When tested by pricking with a pin there was found a total loss of sensation in the foot and for about two-thirds of the way up the leg, extending further up in front than behind. The exact cause of the paralysis was not made out, but the faradic current was employed as a speculative means of treatment. It was used on four consecutive days. A few hours after the first application he could move his toes, and there was partial return of sensation. On the next day he could move his ankle and also use his foot in walking with the aid of a stick. On the third he relinquished his stick and walked unsupported, although his gait was somewhat unsteady. On the fourth he walked like anybody else and left the hospital.

A few months later he returned, telling us that, after leaving the Infirmary, his leg remained perfectly strong for about a fortnight, and he was able to walk without crutch or stick, but after that time it relapsed into its old condition and remained so.—It was just as helpless as before, but the anæsthetic area was now limited to the foot and did not extend up the leg. The faradic current being used a few times the power again returned, and the patient insisted on going home. It was found that he took epileptic fits occasionally. After a second interval of a few months he returned with the leg once more in the same powerless state, but the anæsthetic area smaller, being limited to a patch over the outer malleolus. He was most emphatic in his desire that something effectual should be done, so I cut down upon the popliteal nerves, and picked each up separately and stretched it. In forty-eight hours sensation was everywhere perfect and he could move the foot and ankle. The wound, though slow in healing, was well in about three weeks, and he again left with sensation perfect and motion good. In September, 1882, he came again, telling us that for a month after his discharge he has been quite well, and that one day while walking on the street the power suddenly left it and it relapsed into its old condition. He was seen again in the end of 1883, with the leg still paralyzed as before. His epileptic fits were pretty frequent and he was very dull in intellect.

The nature of this case puzzled us very much. Here was a complete loss of motion and a partial loss of sensation below the knee, which was twice cured for the time being by a few applications of the faradic current, and once by stretching the popliteal nerves, so that the patient left the Infirmary able to walk, and yet on every occasion the paralysis returned.

HERNIOTOMIES.

No.	Sex.	Age.				
1	Male,	41	Inguinal, ...	Strangulated 56 hours, ...	Herniotomy, combined with radical cure,...	Recovered.
2	Male,	53	Do. ...	Do. 10 do. ...	Do. do. ...	Recovered.
3	Male,	60	Do. ...	Do. 10 do. ...	Bowel gangrenous, from violent taxis before admission. Slit up and stitched to wound,	Died on 4th day from exhaustion.
4	Female,	62	Femoral,...	Do. 102 do. ...	Herniotomy, combined with radical cure,...	Recovered.
5	Male,	56	Do. ...	Do. 35 do. ...	Do. do. ...	Recovered.
6	Female,	56	Do. ...	Do. 10 days, ...	Bowel gangrenous. Slit open and stitched to wound, ...	Recovered.

OPERATIONS FOR RADICAL CURE OF HERNIA.

No.	Sex.	Age.				
7	Female,	27	Ventral,...	Sac partially removed, partially crumpled up. Orifice pulled together with silver stitches—subsequently removed, ...		Failure.
8	Male,	55	Inguinal, ...	Sac removed, and pillars pulled together with silver wire, left in situ, ...		Complete cure.
9	Male,	22	Inguinal, with undescended testis,...	Radical cure, as in preceding case, with removal of undescended testicle,		Complete cure.
10	Male,	22	Do.	Do.		Complete cure.
11	Female,	55	Femoral,...	Removal of sac and contained adherent omentum, ...		Complete cure.
12	Female,	42	Do. ...	do. ...		Partial cure.
13	Female,	32	Do. ...	do. ...		Partial cure.

OTHER OPERATIONS.

No.	Sex.	Age.				
14	Male,	10	Intestinal obstruction, ...	Laparotomy, ...		Death after 9 hours.
15	Male,	26	Œsophageal stricture, ...	Gastrostomy, ...		Death on 4th day.

One case of umbilical and one of inguinal hernia, with symptoms of strangulation, were reduced by taxis. One case of malignant stricture of the œsophagus treated by bougies. Died 15 months from commencement of disease.

With regard to the seven cases of operation for radical cure of hernia, they do not call for any remark, as they are fully described in a paper on this operation in the *British Medical Journal* for November 11th, 1882. So also are four out of the six cases of herniotomy for strangulation, where the bowel was not merely released and replaced in the abdomen, but the sac was dissected out and cut away, while in the case of inguinal hernia the pillars of the ring were stitched together.

In two cases of strangulation, however, the bowel was gangrenous, and had to be opened and stitched to the lips of the wound. One patient was a great fat old man, with hardly any vitality. His hernia had only been strangulated about ten hours, but it was very tightly nipped, and most unfortunately had been subjected to such a severe pounding before his admission that the scrotum was all black and blue. The combined result of the low vitality, the tight nipping, and the severe pounding, was that even at the end of the short time mentioned, the bowel was gangrenous and had to be opened. He slowly sank, and died about four days after the operation from sheer prostration, and without any signs of peritonitis.

The other patient was quite the reverse of the previous one—a tough, hardy, healthy woman. She had had a femoral hernia for fifteen years, which had gradually attained the size of a small cocoa-nut. She used to support it with napkins, and got about her work in the house after a fashion, and with much difficulty. Suddenly, after a day's washing, strangulation came on, but she steadily refused to have any operation performed for ten days, till just at death's door. Finally assenting, an incision was made over the mass, through inflamed skin, and then an abscess cavity in the cellular tissue was opened into. Beyond this the sac was reached and opened, and a great mass of putrid omentum tumbled out, concealed in which was a knuckle of gangrenous bowel, already perforated in one or two small places. The rotten omentum was cut away, and the bowel opened and stitched to the lips of the incision. The patient recovered, and eventually an artificial anus in the groin resulted, through which nearly the whole of the fæces passed in a fluid form. Three months after the operation attempts were begun with the view of restoring the continuity of the bowel. I made an effort to do this by introducing a thick piece of india-rubber tubing into the opening, and pushing one end up the ascending bowel and the other down the descending. It was fastened by a piece of stout silk, which hung out of the opening, so that it should not become lost. It was calculated that the continuous elastic pressure of the tubing against the projecting spur or eperon would press it back, and so allow the fæces to pass round the corner, without flowing out by the artificial orifice. The tubing was kept in for a week at a time, and was inserted twice or thrice. At the end of seven weeks the patient left the Infirmary with nearly all the fæces passing by the rectum, and only a few drops of a yellowish coloured fluid exuding from the artificial opening, which was now reduced to the condition of a mere sinus. At the end of three months this completely closed. She was seen recently, the hole having been tight for a year and nine months. Her bowels act regularly, and she has no trouble with them whatever. In addition to this, her rupture is radically cured, there is not the least impulse on coughing, she does not wear any truss, and she does abundance of hard work.

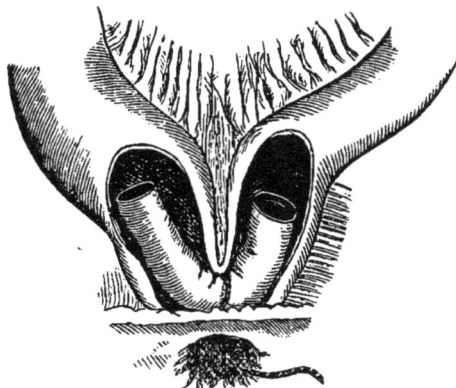

Plan of curing Artificial Anus by the introduction of a rubber tube into the bowel, held in place by a thread passing out of the opening.

In connection with these cases one may note what is the proper treatment when, the sac of a hernia being opened, the bowel is found gangrenous, and obviously unfit to be returned. My own very strong opinion upon the subject is that the bowel should be slit open and stitched to the lips of the wound, and that the ring and surrounding parts should be left absolutely untouched. The

intestine should simply be cut into as if it were an abscess. By
the time matters have come to such a pass that the gut is dead, the
ring will almost certainly be glued around its neck by adhesive
lymph and the abdominal cavity quite shut off from the hernial sac
and its contents. But to this day the surgical authorities say that
the stricture should be divided. The only result of this is that the
protecting barrier, which divides the still aseptic peritoneal cavity
from the putrid sac, is broken down, and putridity spreads upwards
into the abdomen and kills the patient by rapid septicæmic poisoning. Why break down this valuable wall? If it is argued that
unless the stricture is divided the contents of the bowel cannot
escape, then the reply is that experience proves this to be utterly
untrue. In a very short time both flatus and fæces find their
way out. As everyone knows, the nipping of the gut is not produced by a sudden narrowing of the hernial aperture, but by a
swelling of the loop of gut ;—just like the case of the ring slipped
over a finger which cannot be got off again. When the gut is slit
up, its contents are let free and its inflammatory juices escape,
with the result that its swelling goes down and room enough is
soon permitted for wind and fæces to pass, more particularly as the
fæces are invariably quite liquid. Are we arriving at a period
when it will be possible to pull down the bowel, cut the gangrenous
piece away, and stitch the fresh ends together? This, of course,
has been done. I performed the operation myself about six
months ago on a healthy young man, cutting away some eight
inches of gangrenous gut, with a wedge of omentum. The ends of
the bowel were then united and returned into the abdomen, the
sac was cut away, and a radical cure performed. The patient is
now in perfect health, and working as a ship's cook. The question
is not as to its practicability, but as to the following it as a rule of
practice. Possibly it may come to be the regulation thing in well-
appointed hospitals, with abundant assistance; but in private, or in
country practice, where hernias are often cut at night, and with but
scant help, I suspect that the plan of cutting open the bowel and
letting everything alone will hold good for many a day to come.
I have opened the bowel four times. *Case 1* was *in articulo mortis*
almost at the time of operation, and died within a few hours.
Case 2 was the fatal case just recorded. *Case 3* was that of a farm
labourer, in whom about six inches of gut were slit up, crumbled

away, and disappeared. An artificial anus resulted, which closed spontaneously, and he is now well, hearty, and at work, five years after the operation. *Case 4* was the last recorded of the two mentioned above. Thus two died and two recovered, which is not so bad, considering the state of a patient whose bowel has been allowed to become gangrenous.

Next, as to the treatment of artificial anus. After the usual attempts to close the hole by pressure, caustics, and other mild means have failed, Dupuytren's enterotome has been our sole resource. But the remarkable power of elastic compression is known to everyone, and the success which attended the steady pressure of the rubber tube against the projecting spur in the case last described, was eminently satisfactory. I don't know that this is a recognised method of removing the obstruction, and, in consequence, am all the more pleased at its success, and mean to try it again on the first similar case. By the way, on the last introduction of the tube the hole contracted so much that it was very difficult to pull it out. I had reckoned on this, however, and had intended, in case of finding it impossible to remove it without again greatly dilating the orifice, simply to cut it adrift and let it find its way down the gut as best it could. It would not have done any harm.

An Unsuccessful Case of Gastrostomy.—The patient was a young man, aged 26, a barman by occupation. Seven weeks before his admission he accidentally swallowed some sulphuric acid from a bottle which he thought contained rum. As a result, a most intense stricturing of the œsophagus, just above the diaphragm, took place. When admitted, he was utterly unable to swallow even milk, and was greatly emaciated and prostrated. Not the smallest instrument could be got into the stomach for two days, and then a No. 9 catheter was got in, to the upper end of which a wider one was fixed, and so a good supply of milk was thrown in with a syringe. It would be useless to detail the history of the patient at any length during the next ten months. It is sufficient to say that although every conceivable effort with every kind of instrument was made, no real dilatation was effected. The stricture, in addition to being very narrow, seemed also to be very long and tortuous. At intervals the patient became utterly blocked up and could swallow nothing, and during these periods no instrument could be passed.   Three of them lasted for ten, nine, and eight days respectively, during which time not a drop of anything reached the stomach.   He suffered at these times intensely from thirst—more so, indeed, than from hunger.   We kept him coiled up in bed, and warmly covered, so as to keep up his heat, and he got morphia subcutaneously at regular intervals.   By this means as little waste as possible was incurred, but his rapid emaciation was most painful to witness.   At last a No. 8 catheter would be got in, and a pint or two of milk pumped down it, to the poor fellow's intense delight. For the most part, he lived on milk and eggs, and kept in very excellent condition, as he rapidly gained flesh so soon as the catheter could be introduced.   He was an out-patient, and used to come to have his bougie passed at regular intervals; while during his starvation attacks he was taken into the house.   It became clear, however, that the man could not go on this way for ever, and so plans were thought of for forcibly dilating the stricture.   It was obvious that no cutting instrument could be used with safety, first, on account of the great distance of the stricture from the mouth, and secondly, because we had made out that the stricture lay just in front of the aorta, the pulsations of which were communicated to any instrument that was got through it.   I bethought myself of laminaria, and having procured a laminaria bougie, I cut off about three inches of it, and so arranged it in connection with his usual bougie that I was able to introduce it into the stricture and leave it there.   A string was attached to the top of it, and was left hanging out of his mouth, so as to enable us to pull it up if necessary. Two or three times the laminaria bougie was inserted, but after it had been in for a few hours it was expelled upwards—how we did not know.   At last a piece was got firmly in, and remained for two days. During that time he suffered a good deal of pain, and, of course, was unable to get anything into the stomach.   He used to solace himself by filling his œsophagus with fluid and throwing it up again, which relieved his thirst somewhat.   It was then deemed right to pull the expanded laminaria up again, when, somewhat to our dismay, the string broke and it remained in the stricture.   At the end of the 50th hour, however, the patient having filled the œsophagus with some tea, the bougie was felt to descend into the stomach.   For twenty-four hours the results were admirable, the patient being able to swallow soft solids, such as he had never partaken of since the original accident. But after that the stricture closed up worse than ever, and in a few days became so completely occluded that a starvation fit lasting five days ensued. The laminaria was evidently of no avail. A period of two months next elapsed, during which the stricture grew rather worse than better, so that the power of voluntarily swallowing fluids was almost lost, and his feeding had to be done by pumping in milk, introduced by a No. 5 catheter, three times a day. Of this the patient became thoroughly wearied, and expressed his willingness to have the stomach opened. I accordingly performed gastrostomy on February 4th, 1880, under ether and antiseptics. An incision about three inches in length was made parallel to the left costal arch, and the muscles of the abdominal wall being divided, the peritoneum bulged up. This was opened, and the stomach, which lay immediately beneath, was drawn to the surface. It was secured to the abdominal wall by eight carbolized silk stitches, which did not penetrate through mucous membrane, but only through the peritoneal and muscular coats of the organ. The whole proceeding was of the easiest and simplest description. The patient passed a pretty good afternoon and night, being fed by nutrient enemata, and next day the wound was dressed and found all right. On the same evening he was restless, and had a cough and some pain in the chest. A hypodermic injection said to contain a quarter of a grain of morphia was given. Whether this contained more than the proper amount, or whether an amount, to which the patient had been quite accustomed previously, operated upon him in his then condition with unusual activity, cannot be told. Suffice it to say that he was taken very shortly with all the symptoms of opium poisoning, and that the most energetic treatment, pursued during the whole night, was required to keep him alive. He was quite unconscious for thirteen hours. The result of the mauling about to which he was necessarily subjected in the course of the efforts to restore breathing—such as Sylvester's artificial respiration plan—were too much for him, and he died on the fourth day after the operation, delirious and with impeded respiration. At the autopsy, it was found that the wound was in perfect condition, and that the stomach was accurately glued to the abdominal wall. There was no peritonitis whatever. Both lungs were in a pneumonic condition, and in the left was a large patch of grey hepatization. He died from pneumonia, aggravated by an overdose of morphia. The stricture was about three and a half inches long, and corresponded exactly with the view formed of it during life. I cannot quite say what caused the pneumonia, unless it was cold in the theatre during the time of operation, but I fully believe he would have got over the attack but for the unfortunate mishap with the morphia. It was most distressing to lose the patient in this way, while the local conditions were so satisfactory and seeing that he was one in whom permanent benefit and prolongation of life might reasonably have been expected from the operation.

Since the above case I have performed gastrostomy once for malignant stricture of the œsophagus, the patient being in an almost moribund state from starvation. In this case I made a small hole in the stomach at once to allow of feeding, but the patient quietly sank on the second day. Two or three similar cases have been done in Liverpool, but the result is always the same. From what I have seen and read I cannot recommend the operation in malignant stricture. So long as a man can swallow milk and strong beef tea through his stricture, he is as well off as if he were getting it through a hole in his stomach. When he can no longer do this, his time has come. It is far more merciful to give him plenty of morphia and let him die in peace, than torment him with a surgical operation which can only preserve life for a brief period under the most distressing circumstances to himself and all around him.

Intestinal Obstruction—Laparotomy—Death.—The patient was a boy, aged 10, who, until his fatal attack, had never had any illness nor history of bowel mischief whatever. Indeed, his bowels always acted quite regularly. Suddenly there came pain in the umbilical region and lower portion of the abdomen, with obstruction to the passage of wind and fæces. In a few days stercoraceous vomiting came on. He remained for thirteen days with the abdomen distended, and unable to retain anything but the smallest quantity of beef-tea and milk. The only special point revealed by physical examination was a slight tympanitic area in the right iliac region, surrounded by considerable dulness, but what that exactly imported could not be made out. The most accepted view was that there was an intussusception of the small intestine. On the thirteenth day his case seemed so hopeless as regards relief—every method of treatment having failed—that the abdomen was opened by an incision reaching from the umbilicus to one inch above the symphysis pubis. On opening the peritoneum a coil of small intestine at once bulged through the wound. It was intensely congested, and about three times the natural size, being distended with air and fluid fæces. On following this up we came to a distinct twist or volvulus, the heavy intestine having turned over upon itself in such a way as quite to obstruct its canal, while it seemed to have lost all power of righting itself. The bowel was undone, and the contents of the dilated part were found to pass readily onwards with a little pressure. As it was now concluded that the cause of the obstruction was relieved, the abdominal wall was closed. The patient died about ten hours after the operation. He had two free evacuations, shewing that the bowel obstruction was gone, but he remained in a state of serious collapse all the while, and about two hours before his death had severe pain in the abdomen. At the autopsy it was found that the volvulus was quite relieved, but in the cavity of the pelvis were some coils of small intestine, matted in a mass together, and intensely congested. In the upper abdomen was a certain amount of very recent peritonitis, and also a good deal of an earlier date, which had occurred during the first part of his fortnight's illness, and which had matted the ileum and large intestine a good deal together about the brim of the pelvis. The upper part of the jejunum was very heavy and big, and the lower part of the ileum contained thin fæcal matter. In the right iliac fossa the intestines were intimately glued together, and the omentum was also adherent to them. On pulling them asunder it was found that they enclosed a cavity full of fœtid matter, part of the wall of which was formed by the cœcum and a portion of the vermiform appendix. Examining the latter carefully, it was found to be abnormally wide, and bent upon itself by old adhesions. Near the end was a *small, laminated, hard concretion of fœcal matter, about the size of a pea,* and this end, which was exposed in the artificial cavity just mentioned, was perforated.

The starting point of this boy's illness, therefore, was a hard concretion of fæcal origin, situated in the appendix vermiformis. This perforated the appendix and set up a local peritonitis by which the bowels became glued together and formed a cavity in which an intraperitoneal abscess formed. Then ensued a more general peritonitis, as a result of which a paralysis of the small gut ensued. It became full of fluid fæces, tumbled over and became twisted, and never had power to right itself again. Hence a complete stoppage, The abdomen being opened and this twist relieved, it was concluded that matters were put to rights. But the autopsy shewed this was only a secondary condition following on the malady which originated in the appendix.

When it became an established fact that the abdominal cavity could be opened with safety, great hopes were entertained that nearly all intestinal obstructions, not dependant upon malignant disease, would be remedied. Constricting bands could be divided, intussusceptions could be pulled out, and volvuli could be undone. But experience has not sustained this hope, and the recoveries have been so singularly few compared with the deaths, that the operation has certainly not advanced in favour. There will always be two great stumbling blocks in its way. One is that the nature and cause of the obstruction can hardly ever be made out with absolute certainty; and the other is that a very large number of persons recover after obstruction has persisted for very long periods, and after their cases have been regarded as practically hopeless. Moreover, now that purging and other violent remedies have been given up, and treatment is directed to ensuring repose of the bowel by opium, with the view of letting nature overcome the obstruction, such recoveries are decidedly on the increase. Under such circumstances it is no wonder that laparotomy has but a poor chance. Not that the deaths can in any way be ascribed to the operation *per se*, inasmuch as it is generally performed upon persons already moribund, who only want the slightest shock to kill them outright. Of course, the answer to that is that cases ought to be subjected to operation early, before the patient's powers are exhausted. But here again comes in the acknowledged fact that numbers recover provided they are only let alone, while in very many cases when the operation *has* been done, a condition of matters quite irremediable by it has been discovered. I cannot claim to have had any great experience in intestinal obstruction, but such as it is, it has led me to entertain a very poor opinion of laparotomy as a means of relief. In the two last cases of obstruction which have been under my care, I have not seen my way to operate, and the result of the autopsies has been to display adhesions and mattings of a character which could not possibly have been undone by anything short of pulling the bowels literally to pieces. That persons do occasionally die from obstruction, in whom after death it is found that a solitary constricting band is the sole and absolute cause of trouble, is no doubt true. But I maintain that these cases are very uncommon, and that the impression produced by the sight of one such case is so deep, and the feeling of regret that a chance of cure was missed is so great, that the recollection of very many cases which have either recovered spontaneously or have been found after death to have been irremediable, is lost. The conditions are quite different from those of hernia. There we know that the obstruction is caused invariably in a particular way, that it is hopeless if left to itself, and that it is almost certainly curable if operative measures are resorted to sufficiently early. As a result, the sooner herniotomy is done the better. But here we do not know with certainty what the cause of the obstruction is, the cases are not necessarily hopeless if left to themselves, and the chances of relief by operation are very uncertain. I would not wish it to be thought that I utterly decry the operation. I think there may be a certain number of cases where it ought to be done, but they are limited in number. I am rather trying to apologise for the notorious want of success which has hitherto attended the operation, and to show that this has not been from faults in it, but from the circumstances which of necessity surround it.

In all probability the operation of the future will be ileo-laparotomy, where a small abdominal incision is made, and the first piece of distended intestine brought to the surface and opened. This is not a dangerous proceeding; it gives time for nature to act, if she can; and the resulting artificial anus is readily curable if the patient recovers.

DISEASES OF THE RECTUM.

Fistula in ano, ...11 cases.
Hæmorrhoids—clamp and cautery,............. 4 ,,
      Do.   cure by spontaneous sloughing, 1 case.
Cancer of the rectum and colotomy,............ 3 cases (1 fatal).
Syphilitic (?) stricture of the rectum—
    division, ... 1 case.
Rectal hæmorrhage cured by actual cautery, 1 ,,
Dysentery, ... 1 ,,
Recto-vaginal fistula—operation—failure, ... 1 ,,

Fistula in Ano.—One case presented features of unusual interest. The patient was a ruddy, weather-beaten skipper, aged 52. In 1863 he had an ischio-rectal abscess and probably a fistula, which healed after a few weeks. In 1870 he had a similar attack, and again in 1874. But the fistula resulting from the last attack never healed. He went on going to sea in spite of it for four years till 1878, and during these years the parts around the anus were the site of frequent abscesses, followed by numerous fistulous tracks, till at last in 1878 matters became so bad that he had to give up his occupation. On admission there was dense induration all around the anal orifice, extending beyond the ischial tuberosities on each side, and there were numerous apertures all around the anus, from which pus streamed copiously. The parts were blue and brawny. When the finger was inserted into the rectum, however, no stricturing was found, while the discharge had none of the offensive odour, so characteristic of malignant disease. Under these circumstances the case was diagnosed to be one of ordinary fistula of a very severe character, made worse by prolonged neglect. Some of the more superficial sinuses were at once laid open, and during the following two or three months several more were slit up:—on most occasions without an anæsthetic, as the patient complained of very little pain. The wounds were all carefully plugged and many of them completely healed, but some only partially. An unpromising

feature was that in spite of all sorts of applications the induration never diminished, and the portions of those incisions which did not heal assumed a fungating and tubercular appearance. One of the nodules was shaved off and, being properly hardened, was subjected to microscopic examination, but there were no evidences of epithelioma—nothing but inflammatory tissue. Some six months after the patient's admission, Mr. Parker, being in charge of the ward, made a determined attack upon the disease, ripping up the sinuses and scooping out masses of granulation tissue. The hæmorrhage was very severe, but the patient soon recovered, although it subsequently was rather difficult to make out where the anal orifice actually was. A sort of quasi-healing at isolated points took place during the next two or three months, but the induration and tendency to fungate became gradually worse, and a dribbling diarrhœa set in. From this and from exhaustion the patient gradually sank about nine months after his admission, never having suffered any pain to speak of from the extensive mass of disease. To the last there was no sign of stricture of the gut, nor of enlarged glands in the pelvis or groins. What the exact nature of the disease was I cannot say. It was not epithelioma. The microscopic examination was pretty conclusive upon that point, while the healing of numerous sinuses and the complete absence of any stricture were also against it. Nevertheless, the perfect uselessness of all treatment entitled the disease in one sense to be termed "malignant." It was thought possible that a syphilitic history might underlie the case, and so the patient took mercury and iodide of potassium for some time, but without perceptible benefit. This case was in marked contrast with the second fatal case noted in the list. Here the patient, a man of 65, was admitted with fistula of twelve years standing, with the parts on each side of the anus quite riddled with sinuses; but there was no brawny hardness as in the first mentioned previous case, and the patient, in place of being a robust, ruddy man, was pale and emaciated. The sinuses were slit up and for a while it seemed as if a cure was going to be effected, but it soon became apparent that the infiltration and soddening had been so extensive that the coccyx was dead. This and a good piece of sacrum were removed in a necrosed and spongy state, but the discharge was more than the patient could stand and he died exhausted.

All the other nine cases—some of them very severe—being well cut up, recovered and remained sound.   As to injections, galvanic wires, and india-rubber rings, they will afford abundant material for small type letters and communications to the journals for a long time.   Sensible persons will trust to the bistoury and careful-dressing, with a certainty of cure.

Hæmorrhoids—Clamp and Cautery.—Five severe cases were treated in this way, and with the most satisfactory results.   Whatever objections may have been at first entertained to this operation the introduction of the thermo-cautery has completely done away with them.   The subsequent pain, as compared with that from the ligature, though often severe enough, is by no means so agonising or prolonged.   Not long ago I operated on two severe cases—one by clamp and cautery, and the other by ligature.   Both men were done on the same day and lay side by side, so that we could compare their condition.   In addition to having immensely more pain, the patient, whose piles were tied, was unable to pass water for a week, and had the additional infliction of having to be catheterized. Concerning the clamp, the original small one has always appeared to me to be the best.   Somebody having contrived to make it red-hot by an injudicious use of the cautery, the instrument maker straightway loaded it with ivory, and transformed it into a great, broad, clumsy affair, that might be used for an elephant.   How anybody could heat up the clamp to such an extent as to burn the parts beneath seems difficult to understand, as the first principle of a successful operation is that the cautery shall be just above black heat and not red.   Do patients have stricture after this operation? I know of two cases,—not very bad, it is true, but with distinct narrowing.   But in both instances the operation was overdone, and in my own practice I have not seen any such results.   While healing is going, the daily passing of the finger or a rubber bougie for two or three inches up the gut will effectually prevent any stricture.   There have been three grand strides in the surgery of the rectum:—Percival Pott's enunciation of the true principle of treating fistula in ano ; Syme's treatment of fissure of the anus ; and Smith's operation for piles.

Cancer of the Rectum—Three Cases of Colotomy—One Fatal.— One case occurred in a man of 55.   The symptoms dated from about six months previous to admission, and the cancer was situated as high up as the finger could reach. The patient lived about ten weeks after the operation and after death an enormous mass of cancer filled the whole pelvis. It was noticed that the opening into the descending colon had been made with considerable good luck, for immediately below it the bowel had a distinct mesentery, and could not have been opened without entering the peritoneal cavity. The disease was cylindroma.

The second patient was a woman, aged 31, in whom the disease had been in progress for five months. The patient lived nearly three months after the performance of colotomy. The rectum was much thickened in nearly all its extent, and the peritoneum was everywhere loaded with secondary deposits, even on the surface of the liver. The mesenteric glands were enlarged, and had a peculiar glassy look. This led to a careful microscopic examination of the rectal growth by Mr. Paul, which was clearly demonstrated to be a genuine colloid cancer.

The third case occurred in a young, healthy-looking woman of 27, in whom symptoms had only existed for about three months. In her, as in the other two, the earliest symptoms consisted of a frequent desire to go to stool, with straining and the occasional passing of a little blood or jelly-like substance—described by the patient as a diarrhœa. It would be a great point if text-books, by the way, would dilate a little more upon this early diarrhœa of malignant disease. The student's mind is so fixed upon constipation and obstruction as a result of narrowing of the canal, that he does not grasp the fact that the irritated bowel above the disease churns and champs up the motion and pours secretions over it till it is thin enough to be squeezed through the stricture as diarrhœal fluid. In the present case there was a mass of cancer—probably cylindroma—about four inches up the bowel, but the uterus and bladder seemed quite free. My intention was to have performed colotomy, and when the artificial opening was thoroughly established, to have attempted a removal of the rectum with the diseased mass, getting at it by abdominal section. I do not know whether this has yet been done, but it seems feasible enough. I proceeded to the colotomy, but for the first time in my experience of the operation found great difficulty in finding the bowel. It was at last secured, and a very excellent anus made. Next day, however, the patient was hot and feverish, with a temperature of 102°, pulse

100, no appetite, and great pain down the same side, and in the left leg. The wound seemed all right, but was ordered to be poulticed to soothe the pain. On the third day the temperature was 103°, and the pulse 100. The patient repeatedly vomited coffee-ground material, and the abdomen was swollen and painful. Morphia was given subcutaneously. On the fourth day the temperature was 102°, but the pulse was 120, and the vomiting and tympanitis continued in spite of every effort to subdue them. By night the patient became exhausted and comatose, and died. Most unfortunately, no examination of the body could be obtained, but the wound was opened and it was found that fæculent matter had in some way got along the track among the loose cellular tissue, and set up a putrid suppuration there. Micrococci were found in abundance in the discharge. The patient undoubtedly died from septic peritonitis, probably from an extension of the conditions noted in the outer wound. This is the first fatal case I have had, and was extremely disappointing on account of its frustrating my plan for removal of the rectum. For the patient's sake one can hardly regret it much, as when a cancer of the rectum is bad enough to need colotomy, the patient's life is nothing but misery, either with or without the operation; although I admit the operation prolongs life, and certainly obviates the horrible possibility of a complete block.

Non-Malignant Stricture of the Rectum.—The patient was a married woman, aged 30. She had complained of some difficulty in having her bowels moved for about six years, but at no time was there diarrhœa or bloody mucus. In 1875 she was taken with severe abdominal pains, and came into the Infirmary, where, from her own account, incisions were made in the bowel, and she was instructed how to pass a bougie. This she continued to do for about two years, after which time she experienced such difficulty in introducing the instrument that she gave it up. By degrees, however, difficulty of defæcation and general weakness drove her again to seek assistance, and we found a sort of diaphragm stricture about two inches up the gut, which moved up easily before the finger. There was no surrounding induration of the bowel, nor could any be felt from the vagina. The tip of the finger being lodged in the stricture, it was notched at four points with a probe-pointed bistoury, until two fingers could pass through it. In about a fortnight the patient left, passing a large, red rubber bougie every night, and keeping it in for three minutes. Defæcation was accomplished with perfect ease. After the lapse of eighteen months the patient returned to us, saying that she had been able to pass the bougie for about six months, after which the stricture began to close, and she had to give it up. So the same operation was repeated, the parts being found pretty much as on the last occasion. When the stricture was opened up, however, it was found to be longer in extent than previously—not merely ring-like, but occupying about a quarter of an inch of gut, and its surface distinctly rough. There was a shade of hardness also around it, which was not perceptible before. I am sorry to say that I lost sight of this patient, and cannot say how she fared after the second operation.

It is a great query how far non-malignant stricture of the rectum is due to syphilis. It is certainly more common in women than in men. The cases which I have seen have all been in women, and in women in whom syphilis had actually occurred, or was by no means improbable. The present patient had no distinct history of it, it is true; but she was a strolling actress, separated from her husband, and evidently not of the straitest sect. The French are great believers in the syphilitic origin of nearly all non-malignant strictures, except such as are clearly traceable to the cicatrization of dysenteric ulcers, or to injury to the bowel, sustained in such processes as those of tedious instrumental parturition. A very melancholy feature about them is the tendency to relapse even after division. The patients will not pass their bougies regularly, and, even when they do, the stricture closes upon them, and they cannot get them up. I see that Mr. Harrison Cripps not long ago recommended strongly Verneuil's operation of proctotomy, which practically means dividing bowel and everything from above the stricture down to the tip of the coccyx, and I purpose trying it upon a patient now in my ward, who has relapsed after internal division.

Bleeding from the Rectum cured by the Actual Cautery.— The patient was a female servant, aged 32, who had been losing a considerable quantity of blood from the bowel for some months previous to admission, so that she had lost flesh and had become very anæmic and breathless, with palpitation and loss of appetite. There was a little pain at defæcation, but no protrusion of hæmorrhoids was ever noticed. Two or three attempts were made to induce piles to come down, after the patient was in the ward, but nothing could be extruded, and nothing wrong could be felt by the finger in the rectum. Under ether, and with a good light, a careful examination was made with the speculum, and then two very vascular points were found, which were well cauterized with the thermo-cautery, and no bleeding ever occurred again. Such cases are not common, because, when bleeding is of any standing, hæmorrhoids can almost always be found either protruding at the time of stool, or made to protrude by getting the patient to strain while the anal orifice is well thumbed and opened out by the surgeon. I have seen one case since the above in a male, who was at first sent to a medical ward because nothing could be found amiss with the rectum and there was no pain there. The physicians, however, could not possibly trace the source of the bleeding to any higher point, and so he was sent back to the surgical side. An examination with the speculum showed two points like the tops of raspberries about an inch up the gut, while all around was quite healthy. These being destroyed with the thermo-cautery the bleeding ceased.

As every accoucheur sooner or later invents a vaginal speculum, or a pair of forceps, so it is incumbent upon the surgeon to bring out a new anal speculum, with the result that the number of these instruments is legion. After a fair trial of most of them I find the old-fashioned bivalve instrument, slightly altered in a few particulars, the most generally useful. As usually made, however, it is somewhat inconvenient, being about six inches long, and with the handles at a right angle to the blades. Now, as in ordinary prac-

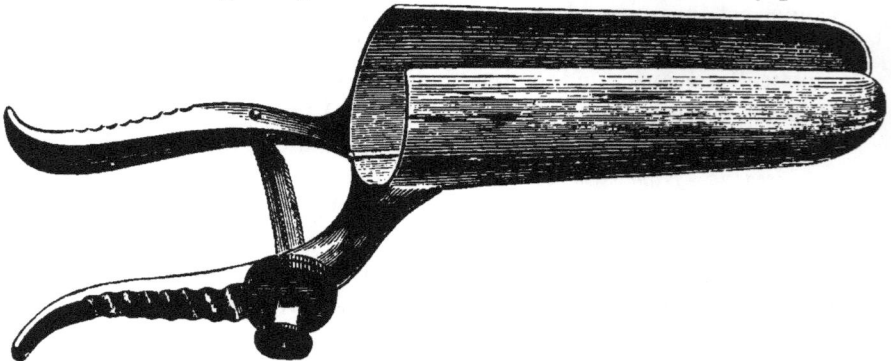

tice the speculum is never introduced higher than about two and a half inches (indeed, it won't go further without causing pain), it follows that the extra protruding part was a great nuisance. I have, therefore, cut down the length, made the point tapering, and thrown the handles almost into a line with the blades, so that the instrument can be turned round in the bowel without their hitching against the ischial tuberosities. The woodcut on the previous page shows the little modifications described.

DISEASES OF THE GENITO-URINARY ORGANS.

	Cases.	Cured.	Died.	Improved.
Traumatic sloughing of the prepuce, ...	2	2	...	...
Cystitis, acute and chronic.	5	3		2
Orchitis, acute and chronic, ...	4	4	...	
Varicocele, excision of portion of spermatic veins, ...	1	1	...	
Vesical calculus—lateral lithotomy, ...	1	1	...	
Do. sounding, ...	1		1	
Perineal calculus, with urinary fistula, ...	1	...	...	1
Intermittent hæmaturia, ...	1	...	...	1
Phymosis and paraphymosis, ...	5	5	...	
Spasmodic retention from cold and drink, ...	1	1		
Incontinence of urine, ...	2	...	...	2
Prostatitis, ...	3	...	...	3
Urinary extravasation—perineal section, ...	3	2	1	
Stricture and perineal abscess—Syme's operation, ...	1	1	...	
Stricture—Holt's operation, ...	13	13	...	
Do. with urinary fistula, ..	3	1	..	2
Urethral fistula—plastic operations, ...	2	1	...	
Hydrocele—tapping and injections, ...	3	3	...	
Vesico-vaginal fistula, ...	1	1	...	
Syphilitic primary symptoms, ...	1	1	...	
Do. ulceration of palate (inherited), ...	1	...	...	1
Do. lip sore, ...	1	1	...	
Do. disease of the testicle,	1	1	...	
Do. sloughing of prepuce and erysipelas,	1	...	1	

Two Cases of Sloughing of the Prepuce and Skin of the Penis from Injury.—Such an occurrence as sloughing of the cutaneous covering of the penis from traumatic causes, apart from any venereal origin, is sufficiently rare. The first patient was a healthy married man, 26 years of age. He was principal manager of a great public-house in one of the lowest and most depraved parts of the city, from which, at closing time, it was not unfrequently necessary to drive away the degraded customers by main force. On one of these occasions a woman was present, who, it was afterwards found out, had acquired quite a celebrity for fighting and who was in the habit, when opposed to a male combatant, of laying hold of him by the privates and hanging on to them. This virago, while being pushed to the door, laid hold of the unfortunate barman by the penis and gave it a severe squeeze. On the following morning the organ was painful and the prepuce swollen. On the second day he took to bed and had medical assistance. On the third day two black spots appeared, which in the course of about six hours spread over nearly the whole of the prepuce, now swollen to the size of a hen's egg. On the fifth evening I saw him and found the whole prepuce and skin of most part of the penis gangrenous, except below. It was slit up and as much as possible cut away, the glans fortunately being found in good condition, although there was a deep ulceration along the upper part of the corona nearly cutting it off. I was afraid lest, when the lower part of the sloughing prepuce separated, the artery of the fraenum might give way, and so recommended his removal to hospital. This was rather fortunate, as on the night of his arrival it was eaten into and smart bleeding ensued. The remaining history of the case is not of much importance. The prepuce and the skin of quite one half of the penis sloughed, but the surface gradually healed over, although the corpora cavernosa were left quite bare. The patient was seen about a year afterwards. The glans had quite recovered itself and the organ was thoroughly useful for all purposes. It looked simply as if a very ruthless circumcision had been practised.

The other patient, a married man of 41, fell from a chair on which he was standing and struck his privates forcibly against the top rail of it as he fell. His prepuce swelled up and soon became gangrenous, together with a considerable portion of the skin of the penis. This sloughed off, leaving a clean, granulating sore which healed in about six weeks. In neither case was there any reasonable suspicion of venereal disease.

H

Varicocele—Excision of a portion of Spermatic Veins. — The patient was a strong, robust man, a small Manx farmer, who, about twelve months before coming to hospital, began to have a troublesome sensation on the left side of the scrotum, not felt in the mornings but towards the end of the day. This soon deepened into a constant dull, dragging pain, which produced such nervousness and depression of spirits that he lost his appetite and had to give up his occupation, as the pain became intolerable so soon as he began to do any vigorous work. From the man's general manner we soon found out that he was one of those unfortunate beings whom a varicocele transforms into the veriest hypochondriac, and whom nothing will improve short of an operation—and not even that sometimes. The pubes and scrotum having been carefully shaved and disinfected, under ether and antiseptics an incision about two inches and a half long was made over the spermatic cord, and the veins were readily isolated. They were tied across with catgut passed beneath them with an aneurism needle at two places, the ligatures being about half an inch from each other. The intervening portions of veins were then cut out. Only two antiseptic dressings were required and in five days the wound was healed. A hard lump remained above and below it indicating that clotting had taken place. In a fortnight he assured us the pain was as bad as ever, and that he could not get any rest with it at night. He went home, but in ten months came back. The spermatic cord of the affected side was perfectly natural, and no more swollen than that of the opposite. The scar of the operation wound was a mere line. He said he had been quite unable to do any work. He had made many attempts, but found that in half an hour or an hour he had such pain that he had to desist. He slept well at night. Nevertheless his life was a misery to him and he came with the express purpose of having the testicle removed. This I, of course, at once refused to do. Hypodermic injections of morphia and various remedies were now tried, but in vain. So in despair I ordered the scrotum and pubes to be painted all over with blistering fluid. The effect of this was so startling that he announced himself much better, and after a second dose he retreated somewhat precipitately to the Isle of Man, professing himself cured. I heard from him a year afterwards saying that he was much better, but far from well. Some days he was apparently all right and others again suffering, so that he did not find himself able to undertake any permanent employment.   Again, more than three years afterwards (Dec., 1883), he wrote, saying "I had very "little pain during the summer, but began to feel worse about the "beginning of October.   I feel very painful now."—Of course the above is one of a class of cases with which every surgeon is familiar, but the reason of the amazing mental depression, which a few congested veins are capable of giving rise to, is at present very inexplicable.   One is very apt to regard the numerous and mysterious complaints of women, who are believed to have obscure affections of the ovaries, with comparatively little sympathy.   But when we find individuals of the stronger sex reduced to the depths of despondency by a very slight ailment implicating the corresponding organs, we must admit that the mysterious pains of so-called hysterical women are not without as good a foundation as those of the victims of varicocele.   That the uneasy feelings and mental distress are quite real I am sure, because I have had to treat two or three medical men for the ailment, and they, with a thorough knowledge of their own complaint, have had great difficulty in keeping themselves cheerful, until the means adopted for their relief proved successful.   As to operative measures there does not seem a necessity for frequently having recourse to them, and personally I have only tried the plan mentioned in the above case. Cutting the veins across with wire loops and threads *may* be all right, but experience has shown that veins so cut across are extremely prone to unite again almost immediately.   So that carrying out Lord Stafford's principle of "Thorough," one would say that the total removal of a good half inch of the whole bunch of veins would be by far the most effectual means.   It does not seem to be in any way dangerous and the wound heals in a few days—at any rate in the eight instances in which I have practised it, such was the case.

Vesical Calculus.—Death after sounding from Uraemia.—G. W., a Welsh quarryman, aged 61, was admitted with symptoms indicative of stone, from which he had suffered for four or five years. A sound was introduced which at once revealed the presence of a calculus.   The patient, however, made a great noise and struggled so much that one was unwilling to frighten him and the sound was withdrawn at once, so soon as it chinked against the stone, without any attempt being made to make out its character or size. Two days later, on a Friday at mid-day, ether was administered and a small lithotrite introduced. With the finger in the rectum it was found that there was a very big prostate and that the stone lay in a deep pouch behind this. It was very difficult to lay hold of, but when caught between the blades, it was made out to be a big stone, very rough and very hard. A diagnosis was therefore made of a large, rough, oxalate calculus lying in a pouch behind an enlarged prostate;—and this in a nervous, thin, and somewhat feeble man. It was decided that lateral lithotomy was the proper thing to do. There was no force whatever used during the examination and no blood was drawn, and therefore there was no ground for thinking that the investigation would hurt him any more than the intro-duction of the sound two days previously. It was found impossible, by the way, to make out the condition of his kidneys by an ex-amination of the urine, as there was a certain amount of chronic cystitis, and the urine contained a good deal of muco-pus coming from the bladder. On the evening (Friday) of the examination the patient's temperature rose to 103°·4. Next morning (Saturday) it was 102°·6. He was kept warm in bed and quinine ordered. No serious apprehensions were entertained. The following morning (Sunday) at 5-30 my house surgeon, being summoned to him, found him in an epileptiform convulsion. He became straightway coma-tose and in six hours was dead. At the autopsy the lungs were found greatly congested and with general emphysema. The heart was sound. The liver weighed 55 oz., and was soft and friable, and so also was the spleen. The bladder, which contained a few ounces of turbid, ammoniacal urine, was moderately hypertrophied. Its mucous membrane was congested and in places ecchymosed, and was evidently the subject of chronic cystitis. Nowhere was its surface abraded and there was no sign of its having been in-jured by the lithotrite. The middle lobe of the prostate was greatly hypertrophied, and projected into the bladder forming a tumour as large as a pigeon's egg, behind which lay a considerable symmetrically branched oxalate calculus. The enlarged prostatic lobe was soft and juicy, as though inflamed, while the lateral lobes were almost normal in size. The right ureter near the bladder was enormously dilated, measuring about $1\frac{3}{4}$ inches across when cut open, but, as it was traced upwards, its calibre diminished. The left ureter in its whole course was about two or three times the normal size. The kidneys weighed about 13 and 14 oz. The right one contained a number of cysts and its pelvis was moderately dilated. There were patches of intense congestion in it, in which were spots of commencing suppuration. The left one resembled the right, and contained a conspicuous number of cysts, but was affected in a somewhat less degree. Both organs were flabby and their structure irregular, owing to old granular change.

From the condition of matters made out after death it was evident that no injury had been done to the bladder or prostate by the examination with the lithotrite. Nevertheless the mere introduction of that instrument resulted in renal inflammation with pyrexia, and, the kidneys being unable to act, the patient, after what was evidently a uraemic convulsion, rapidly died comatose. The serious chronic disease of the kidneys sufficiently explained this. This condition was not capable of being demonstrated beforehand, as the patient had no symptoms of renal disease, and the urine was so turbid and purulent, as a result of chronic cystitis, that no indications as to the state of the kidneys could be drawn from it. That the mere introduction of a bougie into the bladder will give rise in some persons to very grave disturbance of a pyrexial character, and even to death, is well known. In a paper on Urethral Fever, printed in the *Edinburgh Medical Journal* for June, 1871, I recorded some serious instances of this sort, and in explanation of them could only fall back upon the view that they were due to an irritation of the terminal filaments of the sympathetic system, of which the nerve supply of the genito-urinary tract is principally composed. This patient was probably going about with just as much sound kidney tissue as enabled him to live:—the moment there came a shock to the genito-urinary sympathetic system, with its accompanying general pyrexia, they could no longer carry on their work and so the patient succumbed.

The practical question arises, Having detected a stone with the sound, should any further examination or measurement of it be made before the operation for its removal? In a discussion upon lithotomy at the International Congress of 1881, Mr. Teevan and Dr. Otis very strongly argued that this should not be done, but that, a diagnosis of stone being once made, the patient should be anæsthetized, the stone and bladder examined, and the patient cut or crushed on the spot according to the judgment of the surgeon. In the case of those few surgeons whose practice is confined to urinary diseases and whose prolonged experience enables them rapidly to make up their minds as to what particular operation should be adopted, this doctrine may be sound enough. But cases of stone have to be treated by general surgeons all over the country, and an examination with a small lithotrite gives them such information as to the size and character of the stone, as proves of great service in enabling them quietly to determine at leisure what operation is best suited to the patient's circumstances: it being understood that such examination is conducted with the greatest gentleness and delicacy. I do not think that it can be fairly argued that by this the patient loses any extra chance of recovery, because, as in the case just related, if the mere introduction of a lithotrite, which did not even abrade the mucous membrane, served to kill the patient, *multo majus* would a prolonged litholopaxy or a difficult lithotomy do so. So that had he been cut or crushed straightway after the examination he would have died all the same. On the whole, therefore, it may be pretty safely stated that the occurrence of an occasional catastrophe like the present, or of occasional injury to patients, is more than counterbalanced by the certainty and precision with which a general surgeon will attack a stone after he has had time quietly to think over it, to fix upon the most appropriate operation, and to make the best arrangements for safely performing it and meeting any difficulties that may arise during its performance.

Perineal Calculus.—The patient was a man aged 42, who told us that when he was a lad of 15 he fell upon a grid and injured his perineum. His description of his subsequent symptoms pointed very distinctly to a rupture of the urethra, and it appeared that an incision was made in the perineum into which an instrument was introduced. For nearly two years after this a fistula remained, through which a little urine passed, but this eventually closed. He remained perfectly well for twenty-five years, having no trouble whatever with his urethra till six months before his admission. He then had some irritation on micturition, and, examining his perineum he found a hard ridge upon it, pressure upon which produced pain in making water. After a few days a white point came on this ridge, which burst, and an opening formed through which urine came. This caused him such pain that he laid hold of a razor and with it increased the size of the opening, so that at the next micturition, while straining, a stone, of which an exact representation is given in the accompanying woodcut, dropped through

Calculus from the Perineum which came through an incision made by the patient himself.

the opening and fell on the floor.   After a while the orifice almost closed up, leaving behind it a fistulous aperture through which nearly one-third of the urine passed.   For the cure of this he came under treatment.   This was conducted by dilating the urethra with bougies, a slight narrowing opposite the inner end of the sinus being detected.   At a later period a finely pointed thermo-cautery was twice passed along the fistula, and he wore a red rubber catheter for some hours a day, and drew off all his water with one which he introduced himself.   The result was that in about six weeks in place of about a third of the whole amount of water coming through the fistula, only about a twelfth came;—as ascertained by measurement.   He then dropped all treatment and concluded to remain as he was.   He was a miserable, desponding, hypochondriacal patient, and would do nothing for himself unless goaded on to it.   I saw him after the lapse of a year and found a small fistula still persisting through which about five or six drachms of urine would come with each micturition, only he always put his finger on the aperture and so prevented nearly all escape.

There can be little doubt that, as the result of the perineal section performed in his youth some pouched condition of the urethra remained in which the calculus grew.   It was characteristic of the man's morbid mental temperament that, when he found the lump in the perineum, in place of repairing to a medical man, he performed a sort of perineal lithotomy on himself with his razor.

In the second volume of the transactions of the International Congress for 1881, will be found a paper on Perineal Calculi, by Professor Mazzoni of Rome, in which some interesting points are mentioned. One of the chief questions in dispute is whether such a calculus lies in a cyst formed by the dilated urethra, or in one composed of condensed cellular tissue outside that canal. As to its origin it may be a vesical calculus which has stuck in the urethra, ulcerated through it and so got into the surrounding tissue. Or it may start in a perineal urethral fistula or in a congenital urethral diverticulum. When the cyst wall is very large and very thick, Mazzoni recommends that most of it should be cut away.

Phymosis.—Almost every surgeon has his own ideas about the best operation for phymosis. The plan I recommend to students is the following:—Pull the elongated prepuce well out and pass into it a sharp pointed bistoury, taking care it does not go up the urethra. Slit up the prepuce on the under surface on one side of the fraenum. Then with curved scissors snip the fraenum through. The prepuce then falls back over the top of the glans. With the scissors trim round its square corners and then take a rim from the free edge of it. It is necessary to do this because the contraction of the prepuce is always close to its orifice and had best be taken away. By this means a sort of hood or covering is left for the glans, which is very comfortable to the patient. I have seen not a little suffering in boys who have been ruthlessly shorn by the old-fashioned way of pulling down the foreskin and sweeping off as much as possible. A few stitches of fine catgut hold the skin and mucous membrane together, and do not need to be touched. Some kind of subsequent dressing can of course be employed for adults, but for very young children the best thing is to clap over the parts a handful of absorbent wadding well smeared with vaseline.

Three Cases of Rupture of the Urethra and Extravasation of Urine from Stricture. Case 1.—Patient aged 36—History of stricture for eight years. Acute retention for about a fortnight, and rupture of urethra forty-eight hours before admission. Penis, scrotum, lower part of abdomen and perineum all infiltrated, with diffuse cellulitis and suppuration in progress. Perineal section performed, rubber catheter introduced into bladder and incisions made all over the infiltrated parts. Wounds packed for some hours with turpentine to arrest bleeding, and then treated with charcoal poultices. Wounds all cleaned up, but patient's tongue never became moist, and he died at the end of the fourth week from " surgical kidney," suppurative nephritis having completely invaded both organs.

Cases 2 and 3 were respectively aged 48 and 30 years, the former the subject of gonorrheal stricture for four years, the latter of traumatic stricture for three. In neither case had rupture occurred for more than a few hours, but in both the scrotum and penis were enormously distended. The same treatment was adopted as in case No. 1 and both patients recovered.

The origin and progress of extravasation of urine dependant upon stricture are so very much alike in all cases that it is much to be desired that a uniform plan of treatment should be adopted for them. Concerning the necessity for literally slicing up the infiltrated parts there is no doubt. They cannot possibly be cut up too widely or too deeply, and I have frequently made use of a hint given by Mr. Harrison in his Lectures on Urinary Diseases, viz.:— after making the incisions to squeeze the scrotum and penis thoroughly till all the extravasated and putrid urine is pressed out of them. The infiltrated scrotum can be squeezed and emptied like a lemon. There can also be no doubt as to the necessity for opening the urethra from the perineum and passing a soft instrument into the bladder to drain it. The disputable point is this:— having got an instrument into the bladder, through the stricture, should we cut down upon the stricture, divide it and introduce the rubber catheter into the bladder from that aperture:—or, quite neglecting the stricture, should we cut down upon the membranous part of the urethra immediately in front of the anus and introduce the catheter there? My experience induces me to urge the latter plan. Theoretically many arguments may be offered in favour of dividing the stricture as being at once a means of relieving the present distress and of curing the root of the evil. Practically it will be found in many instances to be a very difficult matter. The membranous urethra on the other hand can always be easily reached, if we have succeeded in introducing even the smallest guide: a straight director being then passed from the perineal wound into the bladder, the staff is withdrawn and a rubber catheter easily pushed along its upper surface. The stricture in front, now relieved from the irritation of urine being forced through it, rapidly softens down and becomes amenable to steady dilatation with the bougie.   I have tried both ways, and now, having once put in my guide, I invariably cut straight down upon the membranous urethra, get in the rubber drain and then proceed at leisure to incise the infiltrated parts.   It may be said that it is not always possible to get in a bougie or grooved staff through the stricture.   All I can say is that I have only failed once in doing so, and that was because the scrotum and perineum were so board-like that the instrument could not possible be turned round.   The truth is that in the class of cases under examination there is no complete block-up, leading to frightful distension of the bladder and rupture of the urethra from mere bursting strain.   The history invariably points to peri-urethral inflammatory hardness around the stricture as a first feature.   This softens down into a peri-urethral abscess which is in the act of making its way forwards—creeping quietly towards the scrotum outside the corpus spongiosum—when the thinned urethral wall suddenly ulcerates through, the urine escapes into the abscess cavity and surrounding inflamed tissues, and instantly everything is ablaze.   But during all this time the patient has been getting his water away:—by drops and squirts, it may be, and with infinite agony, but still getting it away.   And, indeed, it is this very fact which so often blinds surgeons to the danger of the hard lump which they can see steadily forming in the perineum. " Why interfere," they say—" the patient is passing water, although with difficulty?"—It being admitted, then, that, even when the urethra gives way, the patient has still some channel, however small, through his stricture, Mr. Syme's dictum, that when water comes out an instrument should go in, comes into play.   So that it may be taken as granted that the cases of extravasation from stricture, where an instrument cannot be got in are very few, and that as a rule the chief difficulty will be found, not from the extreme narrowness of the stricture but from the impediment to the turning downwards of the instrument from the brawny hardness of the perineum and scrotum.—The bleeding from the relieving incisions is generally very sharp for a while.   I always pack them with lint soaked in turpentine, which at once checks any tendency to oozing and disinfects the rotten tissues in the most excellent way.

Stricture of the Urethra.—Seventeen cases were under treatment, of which, one was subjected to external urethrotomy, thirteen to immediate dilatation by Holt's method, and three, which were complicated with perineal fistulæ, to dilatation with bougies.

The treatment of stricture is one upon which every surgeon of reasonable experience is justly entitled to hold his own views. We have at least six methods to choose from:—external division, internal division, gradual dilatation, continuous dilatation by the retention of the bougie, immediate dilatation or rupture, and rapid stretching. The clever surgeon is he who knows which of these operations to select for his particular case.

If the stricture is not a very bad one and not irritable, nothing can beat steady dilatation with the metallic bougie. The one rock ahead is the desire which the hospital surgeon (who must operate *coram publico*) has, even in the present day, "to get into the bladder" at all costs. The unhappy patient being brought into the theatre or waiting room before a crowd of students, the surgeon considers it a point of honour to get something—if only a No. 1—into the bladder. After twenty minutes prodding with all sorts of instruments this No. 1 is finally jammed in; the surgeon triumphs, and the patient is led away bleeding profusely, and possibly with a false passage. A week's rest in bed, with hot fomentations to the perineum, would probably so have softened down this patient's stricture that No. 3 or 4 would have gone in quite easily, to the great facilitating of further treatment.

Of immediate dilatation (Holt's operation) I would desire to speak in the highest terms. I have now performed it about seventy times without any consequences to the patient which gave me anxiety, and with the result of saving much time and pain. The majority of patients have certainly an attack of urethral fever for a day or so after its performance, but this can always be reduced to a very trifling matter, or even staved off altogether, by the use of quinine and opium. On the morning of the operation the patient has two grains of quinine and a grain of opium, and after it two grains of quinine every two hours for perhaps five or six times. A great many of our patients in Liverpool are sailors who have had African malarial fever, and they all declare that an attack of urethral fever is exactly the same thing in all its phases as a malarial attack, only rather sharper. These men *invariably* have the attack after dilatation, while others, who have never been abroad, often escape. The two forms of seizure being so absolutely identical in their phenomena, quinine, the specific for the one, seems the most appropriate remedy for the other.

Of internal urethrotomy, except as applied to the meatus or the penile urethra at a short distance from it, I confess I have a considerable dread. This arises from having seen two thoroughly healthy men die from it in the hands of other surgeons, who certainly would not have died if they had been treated say by gradual dilatation. The operation in each case was quite well done, but the patients died with symptoms of septicæmia. Now this is a very dreadful thing and I would rather have any number of patients wait a little longer for their cure than be subjected to so serious a risk. For, after all, everything that is accomplished by the urethotome can be accomplished by the bougie or Holt's dilator, while to the very worst cases it is not applicable at all, and for them we have to fall back on Syme's external urethrotomy. I have treated a great many strictures, but I never, I am thankful to say, killed anybody by treatment, and I entertain a very strong view that nobody need be killed in the curing of his stricture, seeing that patience will ultimately triumph over all difficulties. I confess I may be somewhat prejudiced against internal ureth-rotomy, but the occurrence of the two catastrophies alluded to certainly shook my belief in it.

To my mind one of the most important advances in the treat-ment of stricture is the recognition of the simple fact, so clearly demonstrated by Otis, that the urethra, except at the orifice, will let a No. 18 pass quite readily; and ought to do so. The former dictum, therefore, that No. 12 was the highest number that ought legitimately to be used is utterly wrong. , But it will take a long time to convince the profession of this. One constantly sees cases of very chronic gleet with urethral irritation—the sure signs of a slight stricture—where one is told that it is not possible that any stricture can exist because No. 12 goes in without difficulty. When such cases are dilated up to 16 or 17 the whole symptoms disappear. One cannot go the whole length with Dr. Otis, who seems to think that everybody who has had a clap should straight-way have his meatus slit up and a No. 20 put in, but there is no doubt that our ideas about the capacity of the urethra have been quite wrong, and that slight strictures may exist, giving rise to distinct symptoms, with a calibre of urethra which was formerly considered quite normal.

What is the result of the treatment of stricture by the various methods just enumerated? As far as my experience goes pretty much the same with all of them in the long run; at all events as regards hospital patients. Within a period of from six months to as many years they nearly all come under treatment again. So that to students I always preach the doctrine that stricture is incurable. Alarming as this may sound, it will be found a much safer doctrine to act upon than the converse. Take a hundred cases of stricture discharged from hospitals and labelled " cured," and see what becomes of them! I will be bound to say that in only a very limited number of cases can further treatment be dispensed with. Therefore they are not "cured." The one thing needful is to instruct the patient how to pass a No. 15 metal bougie for himself, and obtain from him a solemn promise that he will insert it every Sunday morning. With sailors one lays particular stress upon the special day on which the operation is to be done as adding an air of solemnity to the act. To dilate a stricture up to No. 12 and then send the patient out of hospital as a " cured " case is simply a farce, and, as far as students are concerned, a most dangerous and delusive example.

Urethral Fistulæ repaired by operation—Two cases.—The first case occurred in the person of a sailor aged 32 who, five months before admission, contracted a chancre which eat away the floor of the urethra just at the junction of the glans and body of the penis. There was an orifice between a third and a quarter of an inch long, through which all his water passed, spluttering about and giving great annoyance. Through this aperture a red rubber catheter was passed into the bladder. A small opening was then made in the urethra a little way in front of the scrotum and the outer end of the catheter pulled through it. Thus the bladder was drained, and no urine allowed to pass to the fistulous opening. From non-use the urethra in the glans had become almost obliterated. It was dilated with a pair of dressing forceps thrust into it, and then a large metal bougie was passed through it and along the urethra beyond the fistula. The presence of this considerably facilitated the performance of the next part of the operation, which consisted in carefully rawing the edges of the

fistula with a very sharp, small knife and bringing them together
with fine metallic sutures. The edges came together in a transverse line across the neck of the glans. For ten days the rubber
catheter was kept in its place, so that not a drop of water flowed
over the newly united edges of the wound. At the end of that
time the stitches were removed and the rubber catheter also. The
wound was found to have completely united, and the patient passed
his water by the proper channel through the glans. For the next
month the urethra was kept widely open by the passing of metallic
bougies and the fistula formed by the aperture, out of which the
rubber catheter hung, speedily closed. He was seen some months
after the operation with the parts so perfectly natural that it could
hardly be detected that he had ever had anything the matter with
him.

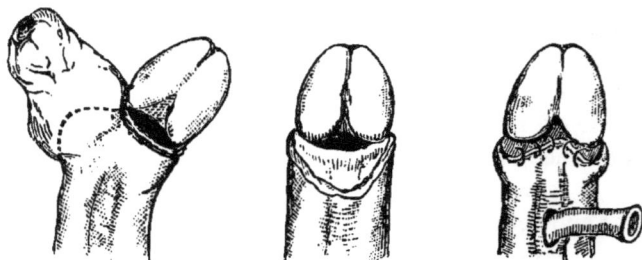

(1.) Glans penis projecting through a hole in the prepuce, with a fistulous opening on the lower surface of the corona.
(2.) Superabundant prepuce removed, leaving only a small flap wherewith to
patch up the hole.
(3.) Fistulous opening rawed and stitched up.—Rubber catheter brought out of
urethra, keeping fistula dry.

The second case was that of a boy about eight years of age who
had a very long prepuce. Outside of this and just behind the
glans he one day tied a piece of string very tightly. The prepuce
in front of this swelled extremely, but the boy, being ashamed of
what he had done, would not tell about it until things had become
unbearable. His mother with some difficulty discovered the string
and cut it through, but by this time it had eaten its way through
the prepuce at one side, and the glans bolted out of the aperture.
This was not a very serious occurrence, but unfortunately the
string had also cut into the urethra on its under surface just at the
junction of glans and corpus spongiosum—in fact in the same site

as in the case last described. The result was a fistula through which all his water passed. When admitted, the elongated hypertrophied prepuce, with the glans sticking out of it, presented a most ludicrous spectacle, looking exactly like a single penis bifurcating near its extremity. I first cut off the enlarged prepuce, but in doing so took care to save a sort of oval flap on the under surface, wherewithal to patch up the fistulous opening. An interval of a few weeks was left for the parts to heal soundly and for this flap to assume its permanent size and appearance. Then the closure was effected. First of all a red rubber catheter was passed into the bladder and brought out of the urethra about an inch behind the glans, so as to keep the wound quite dry. Then a big bougie was introduced to distend the urethra opposite the aperture and the edges of the fistula pared in a $\wedge$ shaped manner. The edges of the flap of prepuce were then shaped so as exactly to fit, and applied against the other. Fine silver sutures were used. After about a fortnight the red rubber catheter was withdrawn and the hole was found closed except two minute apertures at each extremity of the wound, out of which during micturition two tiny jets of urine squirted sideways. Time and the actual cautery closed one of these, and the other was so trifling that I think, by the time the lad attains manhood, it will have diminished to nothing.

Of course in all operations for the cure of urethral fistulæ, which require paring and stitching, a very great deal depends on the patience and dexterity of the operator, but these will avail nothing if the urine is allowed to run over the fresh wound and leak into it. It is of no use keeping a catheter in the bladder with its end coming out at the meatus. The urine runs down by the side of it and gets on to the wound just as badly as ever. The urethra must be tapped behind the fistula and the catheter brought out there. To this I ascribe the success in the cases just mentioned. Of course there is a natural fear lest you should make a fistula at that spot as bad as the one you want to heal, but I believe this to be a perfectly unfounded fear. The reason is that in the fistula which you are attempting to close there has been a loss of tissue (for it is only such fistulæ which ever require operation), whereas in the one made on purpose there is no loss, and it closes up at once, just as any other small fistula does, provided the

calibre of the urethra be maintained at its highest point by the
passage of full sized instruments. Six years ago I encountered a
patient who had lost more than an inch of urethra at the back
part of the perineum, even extending within the rectum (membranous part). I manufactured a new urethra for him and, to prevent the urine flowing over this, tapped the bladder *per rectum* and

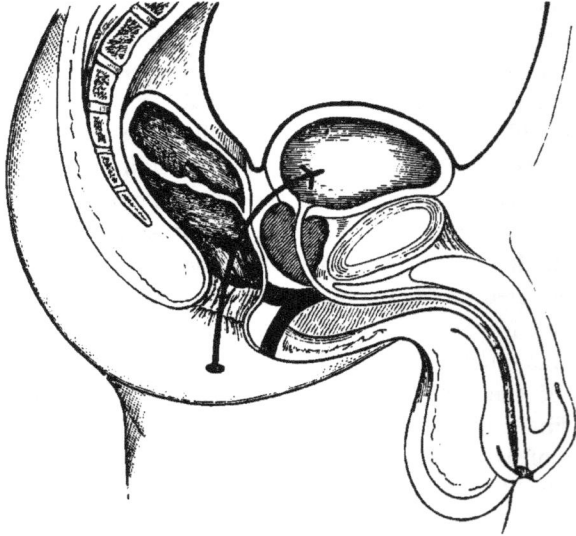

inserted a winged catheter, which drained the viscus and never
allowed a drop of water to pass by the urethra till the wound was
healed. The patient was seen the other day quite well. The
drawing shews the catheter lodged in the bladder from the rectum.
The broad, black line indicates the original fistula which opened
at one end into the membranous urethra and then bifurcating
opened by one orifice into the rectum, and by the other upon the
perineum. Before the patient came under my care the rectal and
perineal orifices were laid into one, as a result of which one long
extensive fistula resulted, running from the perineum into the
bowel. When a bougie was passed into the bladder about an inch
and a quarter of it was visible through deficiency of the urethral
floor.

DISEASES OF JOINTS.

	Cases.	Re-covered	Improved.	Died.
Synovitis of both elbows after scarlatina—anchylosis, ...	1	...	1	...
Synovitis of wrist after scarlatina—anchylosis,	1	1		
Do. hip do. do.	2	2		...
Do. knee (some cases aspirated), ...	10	10		
Acute suppuration of knee after injury—anchylosis, ...	1	1		...
Strumous disease of the knee-joint (no operations), ...	3	2		1
Caries and articular disease of the tarsus (incisions), ...	3	3		...
Acute suppuration of wrist after penetrating cuts—anchylosis, ...	2	2		...
Disease of metacarpo-phalangeal joint of thumb (ultimately excised), ...	1	...	1	
Forcible bending of knee and elbow for anchylosis, ...	2	...	2 slightly.	
Severe sprains of ankle, ...	2	2		
Attempt to restore partial knee dislocation after old disease of the joint, ...	1	...	1	
Morbus coxæ, ...	5	2	1	2

My recollection of the treatment of joint diseases in the Edinburgh Royal Infirmary, when I was a student there more than twenty years ago, is that the whole thing was a farce. There really was no treatment as we understand it nowadays, and the result was that the surgeon groaned aloud at the mere sight of a hip or a knee case. The principle of rest was just beginning to lay hold of men's minds, but they did not completely grasp it. Mysterious virtues were still attributed to the ointment with which the famous Scott's dressing was smeared, while the fact that the real secret of its

I

success lay in pressure and rest was not appreciated. At last,
however, surgeons left off talking about antiphlogistics, absorbents,
deobstruents and discutients, and directed their attention to finding out a mechanical apparatus, which would give a joint real,
absolute, utter immobility. We in Liverpool think there is only
one means—and none other—of obtaining such immobility, and
that is by Thomas's splints. This is certainly not out of any local
personal feeling in favour of our townsman, for Mr. Thomas met
with no little opposition when he first brought out his apparatus.
In a cold, damp climate like that of Liverpool joint diseases are of
universal prevalence in every rank of society, and the surgeons of
this city have in all conscience abundant experience of them—
and to spare. It may be safely said that there is not one who
does not use and appreciate the value of Thomas's splints for the
hip, knee and wrist, and they are in universal application in every
hospital in the town. It is a great pity that their immense value
is not more extensively known, for they have only to receive a fair
trial to secure favour. In a little paper on Synovitis of the Knee
Joint in the Liverpool Medico-Chirurgical Journal for July, 1881,
I said " It is a great pleasure to me to give my testimony to the
"incalculable value of the knee-splint in all injuries and diseases
"of the lower limb below the hip-joint, for which steadiness and
"rest are necessary. Like every first-rate invention it is simplicity
"itself—so simple that one wonders how, with generations of
"surgeons for centuries inventing splints innumerable, the idea
"never occurred till recently of making the patient walk upon his
"pelvis. I consider this splint one of the most important additions
"to the surgical armamentarium that has been made in my time;—
"one, I mean, of the few really useful, and permanent additions.
"For five out of six new instruments seem to me to be devised by
"the man-milliners of our profession for the benefit of that not in-
"considerable section of professed surgeons, who have eyes but
"they see not, and fingers but they feel not. I have noticed with
"regret that the introduction of this splint into other towns than
"Liverpool has been somewhat slow, but I do not hesitate to
"prophecy that before many years its use will be general. But
"—(even Thomas's knee-splint has a "but")—the defect is that you
"must have a splint for nearly every patient, seeing that there are
"few patients whose legs are the same length and whose thighs are

"the same girth. In a large hospital, where numerous splints are
"kept in stock, this defect is not felt. You can always pick one
"out of the lot that will fit your patient; but in private this can-
"not be done, and the patient must be measured and the splint
"made before the much needed support is obtained. Indeed, in
"country places where there are no artificers in splints, this practi-
"cally interdicts the use of the instrument. At my suggestion
"Mr. Critchley of Upper Pitt Street has constructed one, the limbs
"of which can be lengthened or shortened, separated or approxi-
"mated, while the circumference of the thigh ring can be enlarged
"or diminished. One such splint will serve for any adult of the
"average dimensions of leg, and will, therefore, probably prove of
"service to gentlemen in country practice or at a distance from an
"instrument maker."

The grand feature in Thomas's hip and knee splints is that they
allow the patient to go about while his diseased joint is still kept
absolutely motionless. The cure of a pronounced case of hip or
knee disease without suppuration is a matter of twelve months to
two years. But children, who are shut up in bed or chained to
the couch for such long periods, fade and wither away. Their
vitality oozes out of them, inflammatory products break down and
soften, suppuration occurs, the joint is opened and a perpetual
drain sets in, till the unhappy little patients literally melt away
before our eyes. Now much of this is due to bed. Not a few
times I have seen children sent home from hospital to die. The
hospital was sick of them and they of it. But after long periods I
have stumbled across those very children, who ought to have died,
and have found them well—with shrunken limbs and stiff joints,
no doubt, but alive and kicking. And the history generally was
that, when the child got home, the mother took all the splints off,
and lapped the joints in cold water bandages or some old woman's
plaster. Then the child began to crawl about the floor or play on

the doorstep; it began to eat and pick up flesh; its sores dried up and its joint got well after a fashion.   Every surgeon of experience has encountered such cases and been told by the triumphant mother how the cold water cloths or green ointment cured her child after all the Infirmary doctors and their apparatus had failed. What is the moral to be got from such cases?   It is that too much bed and too much surgery kill children.   Fix the swollen pulpy knee-joint firmly on a Thomas's splint, mount the lad on a high patten on the other foot and in a fortnight you will find him playing marbles with the other boys while his health and his joint are mending simultaneously.

Another admirable splint used in the Liverpool Infirmary was invented by Mr. Frank Paul, and goes among us by his name. He originally made it of the hoop-iron, employed to pack cotton bales, thrust into thick rubber tubing.   This iron, however, was some-what rigid, and Mr. Critchley of Upper Pitt Street, Liverpool, who sells the splinting, uses a more flexible iron.   The first woodcut shews a piece of the iron in the tubing, which it, of course, flattens out.   The second shews it applied to a case of ankle-joint disease, with sinuses.   It is generally fastened to the limb with strapping, or paraffin or plaster bandages.   Being quite flexible it can be

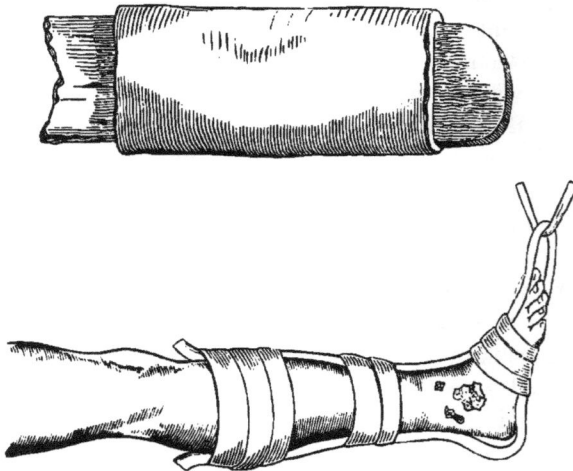

adapted to any surface and yet has rigidity enough to fix a joint. Mr. Paul has lately been employing it in compound fractures of the leg to which it is eminently applicable.

While we may now fairly reckon that all cases of pulpy or gelatinous disease of the synovial membrane are curable in children with time and patience, provided they are taken in hand before destructive suppuration has occurred, there is unfortunately a large number of cases where the disease commences outside the joint and spreads to it, and these are not so amenable to treatment. Our more exact information as to this form of disease is a matter of comparatively recent date; but we now thoroughly recognize the occurrence of inflammatory processes in the epiphysal lines or in the cancellous tissue of the ends of long bones, whence arise abscesses, necroses, or cavities filled with cario-necrotic debris. These eventually reach the joint itself and under the influence of inflammatory action the articular cartilage breaks up by cell proliferation (constituting the condition formerly known as "ulceration of cartilage"), while the joint as a whole becomes the subject of suppurative disorganization. Having been very carefully taught that this ulceration of cartilage was the primary lesion and that the inflammatory mischief subsequently spread to the bones from the joint, I first undeceived myself by preparing about sixteen years ago for the museum of the Liverpool School a number of specimens of knee disease, including parts removed in cases of excision. It was clear that in almost every instance cavities with cario-necrotic debris had opened into the joint. Of course one knows now that such a collection of specimens would be sure to shew a great preponderance of primary bone disease, because specimens of diseased soft parts alone are seldom worth keeping for exhibition purposes. But the effect of this examination demonstrated to me most effectually the very great frequency of epiphysitis or osteomyelitis of the cancellous tissue near the joint, and I have consequently followed ever since with great pleasure the labours of Mr. Thomas Smith, Mr. Croft, Mr. Morrant Baker, Mr. Richardson Cross, Mr. Eve, and others who have paid special attention to this subject. After apparently interminable strifes among the microscopists it is moreover a great relief to find that the investigations of Mr. Macnamara and Mr. Treves seem to have settled the question of the existence of strumous or tubercular deposits in bone—a point which the much despised "practical surgeon" never had any doubt about. As a result of cell hyperplasia, arising in the rigid unyielding tissue of bone, the circulation is cut off from the centre of the diseased area;

its cells, deprived of nutrition become fatty and disintegrate and so a cavity is formed; the osteoblasts and the osseous trabeculæ perish and a strumous abscess results. Or as Mr. Macnamara, in the *British Medical Journal* (October 8th, 1883) holds, agglomerations of cells in the medulla not unfrequently become calcified, ending in the formation of a small, extremely hard nodule in the bone, which acts exactly like a foreign body. This excites inflammation and necrosis of the bone surrounding it, and consequent suppuration. So far the tubercular bacillus has not been found in human bones.

With regard to the knee-joint there can be no doubt that the great majority of inflammations in it arise in the synovial mem-brane, and that epiphysitis and tubercular deposit in the ends of the femur or tibia form the minority. But when they do occur, what is to be done with them? I suspect that, when quite pro-nounced, there is nothing for them but excision.   Some few months ago I performed excision of the knee-joint in a young girl and found three cavities in the lower end of the femur containing cario-necrotic debris, and opening into the joint by small apertures. To have sliced away the femur beyond the cavities would have spoiled the bone, so a moderate slice only was removed and the cavities were then scooped clean out with a Volkmann's spoon, like taking bad bits out of an apple.   A sound limb has resulted; but unhappily this seems the only way of effectually getting at these cavities or patches of rarefying ostitis or deposits of tubercle or whatever the lesion in the cancellous tissue may be.   I am quite aware of the various projects for attacking the ends of bones from the outside by making deep incisions into them or by scooping them out or by boring into them in various ways, all with the view of giving vent to inflammatory products without opening the joint. But the difficulty I have always felt is the impossibility of striking the inflamed centres with anything like certainty.   Even in an adult, who can describe sensations fairly accurately, it is often im-possible to make out with certainty whether it is the tibia or the femur which is affected, so diffused is the pain. How, then, in children are we to hit upon cavities or inflamed areas, which may be lying either just beneath the cartilage or an inch away from it? Of course one does not mean such remarks to apply to well marked cases of abscess in the head of the tibia, where free external opening is often absolutely curative, if mischief in the interior of the joint has not already been set up.

In the ankle-joint proper I have noticed a good many cases of epiphysitis specially affecting the neighbourhood of the inner malleolus, and have several times removed the malleolus bodily (thus opening the articulation) with the result of obtaining a complete cure and a movable joint. Such a case is one of those included in the catalogue of necrosis of the tibia. The child was a poor, weakly thing, nine years old, who about a year before admission got a kick over the inner ankle. After a while abscesses formed round the joint and burst. She came to the Infirmary with sinuses below both malleoli and near the tendo Achillis. Enlarging one of these sinuses over the inner malleolus, I laid bare the joint and found the upper surface of the astragalus still covered with healthy cartilage, while the lower surface of the tibia and its malleolus were in a state of cario-necrosis. I scooped all this away with a spoon and put the parts in proper position again. All healed up satisfactorily, leaving a stiff joint.

As regards the hip there seems to be little doubt that the primary disease there is almost always an ostitis of the upper epiphysal end of the femur. Sometimes this dies *en masse* and is found in the joint as a loose, foreign body.

A consideration of such points only makes one the more anxious that the treatment of diseased joints should be commenced at the earliest opportunity. The early limping and lameness, which are so constantly put down to growing pains and bad habits, these are too often simply the signs of bone inflammation, which, as Mr. Symonds pointed out at the last meeting of the British Medical Association, does not distend the bone or give a swollen feeling so long as it is central and does not come too near the periosteal surface. Rest—absolute rest—in the early stages, with soothing fomentations is the only chance. I think I have seen good come from the use of the actual cautery. It was Syme's great remedy for what was called disease of the cartilages, which we now know means ostitis of the epiphysal ends of the bones. Concerning its power of relieving pain that is beyond doubt, and, seeing the value that veterinary surgeons attach to its use, it may possibly come into vogue again for human beings. One plan of treating joints, which came somewhat into fashion a few years ago, I must positively decry:—the so called "open method." I have seen it extensively employed, and unfortunately employed it myself for a short time, until I discovered that it was the most useless, painful, and fatal form of treatment I had yet tried.

PERIOSTITIS AND NECROSIS.

Necrosis of the clavicle,	...	1 case.
Do.	ulna, ...	1 ,,
Do.	humerus, ...	1 ,,
Do.	jaw from bad teeth, ...	4 cases.
Do.	jaw from injury, ...	1 case.
Do.	femur, ...	1 ,,
Do.	tibia, ...	5 cases.
Do.	calcaneum,	2 ,,
Do.	calcaneum and tibia, ...	1 case.
Do.	metatarsal bone, ...	1 ,,

Periostitis of tibia,	... ...	4 cases
Do.	femur, ...	1 case.
Do.	digital phalanx,	1 ,,·

Syphilitic Periostitis—Actual Cautery and Large Doses of Iodide of Potassium.—One of the cases of periostitis was of some interest. It was a case of undoubted syphilitic periostitis of the tibia occurring in a married woman 28 years of age. Having had one child eleven years previously she next had a miscarriage, followed by a rash over the body, sore throat, slight ulceration of the tongue and rheumatic pains. After recovering from this state, she remained well till three years ago when she began to suffer from severe pain in the left tibia, always worst when in bed towards early morning. The left tibia from about two inches below the knee almost to the ankle was considerably thickened and the skin over it was smooth and glassy. So severe and protracted had the pain been that her health was completely undermined by it. She was thin and careworn, with an anxious, haggard expression, as if in constant dread of the approach of her enemy. She had been taking five grains of iodide of potassium for some time and had tried all sorts of local applications but without relief. She was ordered to take twenty grains of the iodide in half a pint of water after each meal, with five minims of liquor morphiæ added to it. Ether was administered, and with the thermo-cautery at a white heat the whole front of the tibia was scored after the fashion of a gridiron. The effect was literally magical. That night she obtained the first night of comfortable and refreshing sleep she had had for many months. She continued the iodide for about three weeks and left the hospital at the end of a month quite well, with all traces of the periostitic thickening gone and quite free from pain.

The case was particularly interesting to the clinical class, as a demonstration of the immediate effect of the actual cautery and big dose of the iodide in syphilitic periostitis. Like bleeding and blistering and setons, the actual cautery has gone out of fashion,— counter irritants generally are at a discount. This is rather a pity, because there are many cases of ostitis and periostitis, accompanied with great pain, where nothing gives relief like the cautery. Five-and-twenty years ago, as I have just said, it was deemed a most valuable means of treatment in what was called acute inflammation of the articular cartilage, and I have seen many a knee-joint fired in those days with the greatest benefit. No doubt, such cases were really instances of ostitis in the cancellous tissue of the articular ends of the femur and tibia, and hence the success of the cautery. But be that as it may, the cautery deserves more attention than it receives, particularly as its application is now deprived of all its terrors, and the subsequent smarting is speedily relieved by cold water bandages. Concerning the iodide, the ordinary four or five grain dose, thrice daily, seems in a large number of cases to be just enough to produce all the ill effects of the drug, and none of its curative ones. For my own part, I incline to think there is nothing between the two grain dose given very frequently, after the homeopathic fashion, and the fifteen or twenty grain dose given thrice daily. The patients who tell you they cannot take the iodide will almost invariably be found to have been trying the medium dose, and will be found almost as invariably capable of taking the big dose, provided it be administered properly—that is to say, after meals largely diluted, and with a little morphia to sooth the gastric irritability which it occasions.

Necrosis of the Tibia—Bone Planting.—Jas. Hampson, a healthy-looking country boy, aged 11, about two months before his admission (January, 1880) got a severe wetting one day. The same evening he was chilly, and then feverish, and soon began to complain of severe pain in the left tibia. He went through all the stages of ostitis, with sub-periosteal suppuration, and after three weeks of poulticing the skin gave way, and the matter escaped by two holes a short way below the knee. In the meantime the inflammation had extended to the knee-joint, so that when he came to hospital we found him in a very weak and emaciated condition, with hectic cheeks and dry, branny skin. The tissues over the tibia were undermined,

and the joint was swollen, glassy, exquisitely tender, and evidently
with some fluid in its interior.   It was bent up at an acute angle.
A free incision was made over the tibia, and shewed a large portion
of shaft evidently dead.   This incision drained the parts, and all
our attention was then given to saving the knee-joint, which seemed
as if on the point of bursting out into a state of acute general inflammation, in which case amputation would have been the only resource.
By absolute fixation on a Thomas's splint, and gentle traction from
the foot with shot-bags over a pulley, combined with soothing
lotions, the bending was overcome, and the joint inflammation
slowly subsided.   About six weeks after admission the dead piece
of tibial shaft was removed.   It was so large that it had to be sawn
across in the middle and each half pulled out separately.   It was
about 4½ inches in length, and involved the whole thickness of the
shaft, except at the very ends.   The periosteal sheath was, of course,
left intact, like a great empty trough.   During the next two months
this slowly commenced to fill with granulation tissue, but the loss
of bone was so considerable that we feared the powers of the periosteum might not be quite equal to repairing it, and it was accordingly determined to attempt bone-planting.   On April 14, a healthy
man required to have amputation of the thigh performed for an
injury.   He and my patient were both etherized at the same
moment.   As soon as the limb was removed, three small pieces of
bone from the femur were nipped off with bone forceps, and a little
strip of periosteum cut off.   Taking them to my patient, I made
four small incisions in the granulation tissue on the sides of the
furrow, and buried the pieces of bone and periosteum in them,
covering them completely over, and pressing the granulations gently
down upon them.   The three pieces of bone were planted on the
outer side of the furrow, and the piece of periosteum on the inner.
For a few days only a bit of gutta-percha tissue was laid over the
wound, and after that it was dressed with absorbent cotton wadding
dipped in boracic lotion.   We watched most carefully to see if the
fragments would reappear, but they never did, so that there can be
no doubt that they lived, as if they had died they would infallibly
have come to the surface.   By degrees the furrow slowly filled up,
and the skin began to creep in over it.   We could plainly make
out that over each piece of planted bone the granulations formed a
kind of hillock, as if there was a centre of activity beneath, and by

the time the parts were sound and ossified it was clear that there was a much more abundant formation of ossific material on the outer side (where the fragments of bone were planted) than on the inner side. At the point where the piece of periosteum was buried no appreciable change was visible. The boy was seen three years afterwards, with a perfectly useful leg, only an elongated depression shewing where the bone had been lost. The outer side of the depression was distinctly higher than the inner. The knee-joint was still somewhat stiff, and will probably always be so.

In reviewing such a case as this, one must not allow one's surgical ardour to ascribe too much importance to the buried bone fragments. In a healthy boy of eleven, if the periosteum has not been killed by intense pus-pressure at an early stage of the complaint, its reparative powers are astonishing. So that I quite believe the periosteal trough in this case would in time have become filled with bone. But the case proves this much at all events—(1) that isolated fragments of bone may be buried in granulation tissue and live; and (2) that they will assist in the formation of new bone. More practical knowledge of the subject of bone-planting is yet wanted before much can be said about its future utility. The best instance of its real value with which I am acquainted is the case recorded by M'Ewen, of Glasgow, where fragments of bone were planted in the arm, and practically restored a lacking piece of humerus.

Necrosis of the Calcaneum.—Two cases of limited necrosis of this bone occurred, one in a man of 62, and the other in a youth of 18. In the former case the piece, or pieces, of dead bone had come away seven years previously, leaving only a sinus. I have noted five cases of limited necrosis of the calcaneum, and have observed that they have all occurred either on the outer or on the inner aspect of the bone, below the malleoli, and just about the attachment of the lateral ligaments of the ankle-joint. The points of insertion of certain powerful tendons or ligaments are admittedly liable to be the seats of necrosis, or cario-necrosis, and probably it is so here. The worst of it is that, even when the sequestrum has come out, there generally remains behind a small carious cavity, which communicates by a sinus with the surface, and obstinately refuses to heal. The only way is to make a very free incision, indeed—a crucial one, if need be—and thoroughly clear out the cavity with a

Volkmann's spoon, following this up with firm packing. And, even then, it is most difficult to induce a sound healing from the bottom to take place. When once the thin, hard, outer skin of a tarsal bone is broken into, and its cancellous tissue opened up, it is very hard to prevent a cario-necrosis from extending further.

SPINAL DISEASES

OF antero-posterior curvature, without suppuration, eight cases were under treatment.—With suppuration, six cases.—Of advanced lateral curvature, two were bad enough for in-door treatment.—One case of sacro-iliac disease.

A sufficient time has now elapsed since Sayre introduced his plaster of Paris jacket to enable us to form a fair judgment as to its merits. As Liverpool was one of the first places in this country where he demonstrated its mode of application, the system has probably been tried as extensively here as anywhere. This has particularly been the case at the Children's Infirmary, where my friend Dr. Oxley's long and extensive practical experience of Sayre's plan certainly entitles him to rank as an authority upon the subject. At first, as is commonly the case, an exaggerated estimate of its value was formed. There were to be no more crooked backs nor psoas abscesses ever again:—and now the pendulum has swung back, and not a few writers have pronounced the whole thing a sham. While avoiding either extreme, I personally regard the plaster jacket as the most important advance yet made in the treatment of antero-posterior curvature:—indeed the only mechanical advance of any moment. As Thomas's splint is to knee-joint disease, so is Sayre's jacket to Pott's curvature: they have raised these maladies from the pit of almost utter hopelessness to within a measurable distance of cure.

The practical application of the jacket as regards its main features does not seem capable of much further improvement. With regard to suspension, however, it is satisfactory to see that the original plan of swinging up the patient, till he twirled round like a roasted apple on a string, is abandoned. Of its painfulness and discomfort there was no doubt:—of its utility in straightening the spine there was the greatest doubt. It is obvious that, had it possessed this power in any degree worth talking about, a great many serious and possibly fatal accidents would by this time have occurred. Almost from the first I have used the swing merely to raise the patient's heels just off the ground, and this not with any idea of straightening the spine, but simply because the chest and abdomen were put into the most natural position for having a close-fitting and comfortable jacket applied around them. That the modified use of the swing has this advantage I strongly maintain. For the same reason, Mr. Davy's plan of laying children in a hammock is a very useful one. It does not alarm the little patients, for one thing, and it puts the body into an easy position for receiving a jacket, for another. Or Dr. Walker's method of building up the jacket with the patient in the recumbent position is, in not a few instances, of great assistance, when the lower extremities are paralyzed.

When one comes impartially to inquire how it is that the plaster jacket has many detractors, there seem to be two reasons. The first is that many people have expected too much from it, and are consequently now undergoing a revulsion of feeling; and the second is that more people don't put the jacket properly on, to begin with. It was expected to work miracles at one time ;—the people who expected this now write letters to the journals calling it names. After all, it is nothing more nor less than a splint to the backbone. What splint in the whole surgical armamentarium is there which will keep a broken bone quiet if the muscles attached to it are allowed to act ? Such an idea is an absurdity, Nevertheless, this is what many people expect from the jacket. The patient with inflamed vertebral bodies and swollen discs is to walk about, is to sit up in a chair, is, in short, to conduct himself just as usual, and yet is to believe that his spine is as rigid as a rod of glass, because, forsooth, he has a shell outside of his skin. It is as impossible as it would be to control the movements of an eel by grasping it with one's hand. I am, therefore, strongly of opinion that in the early or acute stage of Pott's disease it behoves the surgeon to take every ounce of weight off the affected vertebræ, and to procure for the whole column the most absolute rest. There is only one way of accomplishing this, and that is by laying the patient on his back. The plaster jacket then provides him with a rigid and immovable matrix, which, although not equal to the task of overcoming the voluntary movements of the spine, is more than a match for the involuntary ones. Having fitted him with a jacket, I place my patient upon a tray with a mattress in it, by which means he can be transported from bedroom to dining-room, and from dining-room to lawn or sea shore. Or his tray can be put upon a skeleton carriage with wheels, so as to allow him to be taken anywhere. If he be a poor man's child, he must, of course, lie where he can; but he can generally get a change of room once a day. So soon as there is reasonable ground for expecting that the acute stage has passed away, then he gets up and wears his jacket for a year or eighteen months—or, indeed, until it is clear that his vertebræ are once more sound. I know that in insisting upon a period of recumbency as essential I am at variance with many who think that there is no need of such, and that Sayre's jacket from the commencement is all that is required; but I hold that so long as a child's general health keeps good upon the tray, then he is in the best possible position for cure; while readily admitting that he must by no means be allowed to lie for a single day after it is seen that maintenance of the recumbent position is telling upon him.

Concerning the second reason why people make complaints about the jacket, it is simply because they don't put it on properly. All the tales about dirt, pain, sore backs, &c., &c., are quite true, I fully believe; but they do not reflect upon the jacket, but upon the man who puts it on, or the mother of the child who neglects it. Any splint, badly fitted on, will cut a patient, but who would blame the splint? Blame the surgeon, if you like. The simple disproof of all these objections is that the surgeons who have had the greatest experience with the plaster jacket (and who are those who always take the most pains) do not get these misfortunes with it. I have seen children brought to the Infirmary with what were called jackets on them, which amounted to little more than loose wisps of plaster bandage about the loins. As to that support which a well-fitted jacket gives, throwing as it does the weight off the bodies on to the articular processes of the vertebræ, they had no more pretensions to it than the usual red flannel binder. Of course, one can quite understand persons who put these jackets on not quite realizing the immediate cure of spinal disease which they expected, while, by the surgical mechanician and by a certain class of highly mechanical surgeons, it is only natural that so simple and inexpensive a matter as a plaster jacket should be viewed with the profoundest contempt.

As regards lateral curvature, the earlier stages of that complaint

are quite remediable—at all events among the better classes—and
to Mr. Barwell and Mr. Liebreich the greatest praise is due for the
clear way in which they have called the attention of the profession
to the early signs of the complaint, and for the practical items in
treatment which they have described. It has always seemed to me
a singularly unfortunate thing that something more is not done by
us as a body to get into the country a trained class of rubbers and
shampooers, and persons who can do *massage*, such as are found very
commonly in large continental towns. In London, no doubt, they
are to be found, but the great provincial towns are very badly off
in this respect. Had we trained persons of this kind to fall back
upon, the tribe of genteel quack rubbers who trade upon the
credulity of rich persons with families of young daughters would
be exterminated. That they do good by their rubbing I have no
doubt in the least, and the sooner we leave off simply abusing them
and take to the employment of what is good in their system the
better. Our eyes have been opened of late years to much of the
so-called secrets of bone-setters, and great good has thereby accrued,
and fewer persons in the future will have to resort to these individuals. The great plan in such matters is never to despise one's
adversary, but to find out the secret of his strength and rob him of
it. There is a great deal to be done for many troublesome, although
not deadly, complaints by properly regulated friction and movement
of muscles and joints. It is much to be hoped that the Swedish
movement-cure will be taken up and made a legitimate part of our
remedial armamentarium, and not sneered at and decried until it
becomes a mere source of quackery. The way to prevent its becoming such is for the profession to patronize it and its professors,
and so render them dependant upon us, and anxious to do what we
wish in aid of our other treatment. Taking it all round, however,
lateral curvature may be considered a disease of the rich and not of
the poor, and consequently the cases of it which one sees in hospital
practice are usually the very aggravated ones, which have rendered
the patients quite useless for work. Two such cases were under
treatment, the amount of deformity in each being excessive.

DEFORMITIES OF BONES.

Genu Valgum—Ogston's operation—1 case.
    Do.   M'Ewen's operation—4 cases.
Bowed Legs—forcible fracture of thigh and leg bones—2 cases.
Subcutaneous osteotomy for badly united fractures, 4 cases, with one death in a diabetic subject.
Breaking down adhesions, &c., of badly set fractures—1 case.
Repair of injury to the Nose—1 case.
Congenital Deformity of the Pelvis and Lower Limbs.

The operations for genu valgum have now been so popularized that nothing need be said about them.—Personally I prefer M'Ewen's method as being simple, free from danger and accomplishing its end as well as any other. But it may not be amiss to say a word about the after treatment, which is accomplished in Liverpool with the greatest success by means of the Thomas's knee-splint. The great object is, if possible, to turn the knock-knee into a mild condition of bow-leg for the time being. As the limb lies between the two rods of the splint, it is kept pulled down to the foot-piece by means of elastic extension and so steadied. With an elastic bandage the upper part of the femur can then be kept pulled out to the outer rod, while the knee and leg are pulled in to the inner. This can be done so that the position of the bones can be maintained with the greatest accuracy, while such an amount of movement of the limbs in bed may be permitted to the child as prevents the confinement becoming unnecessarily irksome. The woodcut, taken from a photograph, shews this method of treatment. The legs below the knees are seen to be lashed to the inner limbs of the splint, while above the knees broad elastic webbing bands, fixed to the outer limbs of the splint, exercise a constant outward traction. The lithographs shew a case of successful treatment, where the boy was a hopeless cripple and unable even to stand. The first drawing shews him supported by the hands of a nurse in order to be photographed. After the osteotomy he had as good use of his limbs as any other healthy lad.

K

Concerning the age at which operative proceedings need be resorted to, I cannot help thinking that about five is the earliest at
which it need be done, and that before that age a cure should be
effected by mechanical means.   Indeed, whatever may be the case

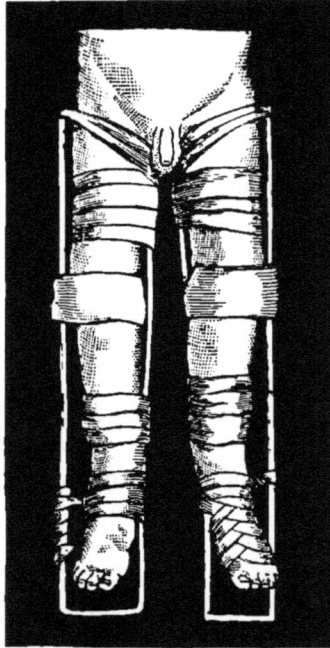

with hospital patients, it is certain that this is true of private ones
who can give time, money and attention to ensure a cure.—In
their instance the limbs are very accurately measured for a Thomas's
calliper splint, the lower part of which fits into the sole of the
boot.   By a rack and screw arrangement the rods of the splint can
be so lengthened or shortened that the weight of the body is taken
off the knees and mainly borne by the pelvis, while the bones
meantime are drawn to one or other side by elastic bands.   During
the whole process of cure the child can be allowed to walk about
with its splints on—an inestimable advantage to the little patients.—
Indeed, by this means knock-knee up to twelve or fourteen years
of age may be remedied, with time and patience.

Bowed Legs—forcible fracture.—This is a most admirable proceeding for hospital patients, who cannot be sufficiently looked

after in the way of splints and rest. Small children's bones can be readily fractured, when they are anæsthetized, by a strong hand, although , it is often surprising how tough they are and what an amount of force is required to fracture them. For bad bow-legs I have done the operation in a good many cases during the past seven years with excellent results, breaking the femora and tibiæ at once on both sides and putting the limbs up immediately in plaster of Paris. As there is no contusion or bruising of the soft parts there is no fear of swelling, and so the plaster may be put on quite safely as soon as the operation is done. The whole process both sounds and looks very horrible, but the children do not seem to suffer any pain afterwards and, as the fractures are only simple ones, it seems practically devoid of danger.

Subcutaneous Osteotomy for badly-set fractures was done four times, once with a fatal result. In the case of David H., aged 39, the femur was the seat of injury. He was captain of a ship, and fell 85 feet upon deck, from the fore-topsail yard, while at sea. He broke the right femur all to pieces in the neighbourhood of the great trochanter. There was no surgeon on board, but he improvised a kind of splint for himself, and lay quiet in his bunk for 135 days, during which the voyage lasted. The bone knitted firmly; but, unfortunately, owing to the position in which he lay in his bunk, the shaft below the point of fracture got slewed round, and so united to the upper fragment in such a way that the whole lower extremity was permanently everted. His progression was consequently most remarkable, as he walked with the inner edge of the right foot forwards and the toes pointing directly outwards. His attempts at bending the knee were also most peculiar, as owing to the twisting round of the femur, the normal lower attachments of the thigh muscles were all altered, and they would not work properly. The femur was sawn subcutaneously just below the trochanters about two-thirds through, and then broken. The upper and lower fragments being then brought into proper line, the limb was put up on a long splint until the small saw wound had quite healed. After that he was fixed in a plaster case which enveloped the pelvis and thigh down to the knee. He made an excellent recovery, and was discharged with the one leg as good as the other, with the exception of the shortening which had occurred during the healing of the original fracture. As he was quite useless for sailor's work previously, this was an inestimable boon to him.

A girl of 15 sustained a fracture of the bones of the leg from a kick about two years before her admission. They united at an antero-posterior angle, in such a way that she was quite lame, and could only walk with fatigue and difficulty. Subcutaneous osteotomy was performed, the leg being treated on a Thomas's knee-splint, with an elastic bandage pulling the shaft of the tibia backwards, and another pulling the heel forwards. When last seen she was walking about free from all lameness.

One case was most unfortunate in its result. The patient sustained a severe compound fracture of the right leg six years previously, and nearly died from pyæmia. After many months' residence in hospital he recovered; but, beginning to walk rather soon, the callus at the seat of fracture yielded, and the bones became bent at a very sharp angle about six inches above the ankle. In this position they at last firmed up, but the limb remained so useless that he eventually came to the Infirmary for the purpose of having it amputated below the knee. I imagined, however, that the leg might be brought straight, and attempted to do so. Under antiseptic precautions, I sawed a wedge out of the tibia opposite the angle, and then divided the fibula. This was a matter of much difficulty, as the bones were greatly thickened and as hard as ivory; and even when it was accomplished I could not bring them into line, owing to the contraction of the muscles of the calf, and the amount of cicatricial tissue resulting from the old injury. The result of the operation, therefore, was not satisfactory as regarded the immediate remedying of the deformity, but I trusted that time and elastic extension would gradually bring the bones into line. On the day following the operation the patient was very sick and prostrate. On the second day this state of matters still continued. On the third day the pulse was 108°, and the temperature 98·2. The tongue was brown and furred, and the vomiting still continued. By the afternoon of the same day he was very bad, the breathing being laboured, the lips flapping, the skin cold, and the pupils dilated. He could not pass his urine, which was accordingly drawn off, and found to contain a slight trace of albumen, *but a considerable quantity of sugar.* He continued restless and even delirious during the night, but could be roused to consciousness by loud talking. His temperature fell to 92°. On the fourth day he died, the temperature suddenly rising to 98° about an hour before death. There was no putrefactive condition of the wound to account for death, so that we were compelled to fall back upon the glycosuria to account for it. That his condition was not one of pyrexia was shown by the fact that his temperature never rose above the normal point, and after the second day was invariably below it, till within an hour of his death. Speaking broadly, his symptoms were "cerebral," so that we came to the conclusion that he was poisoned in some way, almost certainly by reason of the presence of sugar in the urine. This secretion had, unfortunately, not been examined before the operation, as the man had a thoroughly healthy appearance. It only shows how a neglect of this precaution may lead to serious results. I should certainly not have done the operation I attempted had I known him to be diabetic, but would have persuaded him to let matters alone, or, at the most, to have had the less formidable operation of amputation performed. What part could the prolonged administration of ether during the operation have played in this case?

DEFORMITIES (SOFT TISSUES).

Diseases.	Cases.	Cured.	Improved.
Contraction of the palmar fascia, ...	4	4	0
Talipes equinus (infantile paralysis),	1	0	1
Torticollis (division of st. mastoid),	1	1	0
Cleft palate, ...	2	1	1 (failure).
Spastic spinal paralysis (tenotomy for contracted tendons), ...	1	0	1

Contraction of the Palmar Fascia.—All four cases were treated by subcutaneous division. A fine tenotomy knife was inserted beneath the skin and, the blade being turned to the horizontal position, the skin was carefully but freely separated from all the subjacent contracted tissues for some distance—a most important part of the proceeding. Then the ray of contracted fascia was divided, and afterwards, if necessary, the flexor tendons. All the patients seemed to have a good deal of pain after the operation. Three out of the four have been seen lately, and although the fingers are not quite as good as when they left the Infirmary, they are nearly so. They are all quite useful and movable in place of being tucked down into the palm of the hand as they formerly were.

Operations for the remedy of deformities are, more than any others, those which require to be tested by time before anything can be said about them. As to sending the patients out of hospital "cured," nothing is easier and nothing more pleasant than to write glowing descriptions of their condition on leaving. Let us see them two or three years afterwards, and this more especially if they are poor people. The rich man takes pains to prevent his deformity recurring by the careful use of instruments, but the poor man pitches all his mechanical apparatus to the winds the minute he leaves the hospital gates, and never turns up again until the deformity has once more become so serious that he cannot earn his bread on account of it. In the four cases just recorded the results are sufficiently satisfactory to shew how valuable a remedy subcutaneous division is, and how likely to be permanent, seeing that not the slightest care was taken by any of the patients, whereas private patients would have been sure to have worn apparatus for a long time.

As to the cause of the contraction I cannot see that any mere mechanical view will account for it. It occurs in persons of every variety of occupation, including those who do no manual labour at all. Furthermore, the remarkable way in which it affects the little and ring fingers by choice, and the curious tendency to symmetry which it shews are opposed to the mechanical view. The most likely notion is that it is associated with a rheumatic or gouty diathesis, and that consequently (among the rheumatic at any rate), it is somewhat prone to occur in persons who follow out-door manual occupations, such as gardeners and men who handle the pick and shovel.

A Case of Primary Lateral Sclerosis (Spastic Spinal Paralysis.)— The patient was a fine, big lad of 17, a maker of clogs by trade. His trunk and upper works were all sound, but from the age of three months he had been partially paralyzed in his lower limbs, so that he was a cripple. This was his own account, but it might quite well have been a congenital affection. Ross says—"It is "occasionally observed in childhood, which might suggest the "existence of a congenital defect of some part of the spinal cord." He had always to be carried to school, and, even after he had learnt the trade of "clogging," had to be carried to the place where he worked, or had to creep along, clinging to the walls of the houses, until about a year before he came under observation. He had always been able to get about a little, but thought his power of progression had improved somewhat of late. As he lay in bed the legs were not at all wasted or thin, but muscular and well developed. It was seen at the first glance, however, that they were rigid and awkward. The knees were either close together or the legs twisted and locked, and it took a little force to separate them. There was a good deal of permanent flexing at the knee-joints, so that the popliteal space could not be brought down to the bed by about four inches, while the knees themselves had a pointed look from being always bent. The toes were drawn back almost on to the dorsum of the foot, while the heels were pulled up so as to constitute a talipes equinus. The right leg was considerably the worse of the two: neither thigh could be flexed on the abdomen, and when we attempted to do this (or indeed any other movement by force) the muscles resisted, and the limbs trembled violently. His progression was most singular. He walked with the aid of sticks on the fore-part of the feet, with the knees bent and the heels drawn up from the ground. As each step was taken, the one leg was, so to speak, worked round the other, with a sort of scuffling movement, the legs appearing as if they could neither detach themselves from the ground nor from each other. Two or three turns up and down the ward was as much as he could do without a rest, as he then began to have pain in the knees It occurred to us that, if the knees could be straightened out and the heels let down, a great advantage would be obtained for him, and so one tendo Achillis was divided. So soon as he got about and found he could get the heel down to the ground, he desired to have the knee unbent. This was done by dividing the hamstrings, and freely incising the fascia lata just above the back of the knee-joint. Over the biceps tendon a small incision was made, the tendon was displayed and the external popliteal nerve drawn out of the way. I consider this essential to the safety of the nerve if a thorough division of the tendon is to be accomplished. Any dissection will shew that no surgeon, however good an anatomist, can make certain of not injuring the nerve if he operates subcutaneously. Such being the case, the tendon ought to be exposed and divided through an incision, made under antiseptics (as was the case here), a couple of stitches put in the skin, and primary union obtained. So much improved was the patient's gait that in about three months he came back and had the tendo Achillis and hamstrings divided on the left leg. He was seen the other day two years after the operations. He walked with the help of a stick with his knees a good deal bent, but not as much as they used to be, while he could now keep his heels on the ground. He rolled a good deal in his walk, but his legs were apart, and not in contact with each other as formerly, while he had lost a great deal of the old corkscrew, rotatory mode of progression. The muscles were still very rigid, and the thighs were scarcely as large as formerly. Nevertheless, he assured me that he had greatly benefited by the division of the tendons, and had much more walking power in him than of old. For instance, he walked about a week previously a distance of four miles with only one rest. The principle of dividing tendons in spastic spinal paralysis is, of course, no new thing, but the present case adds to the weight of evidence that it is a useful thing to do. By the way, the patient informed me that he never at any time had any sexual desire, and the penis never became erect.

No.	Nature of Injury.	Patient's Age.	Result.
1	Compound fracture of the parietal and temporal bones, ...	16	Death.
2	Compound fracture of the parietal bone — trephining, ...	75	Death.
3	Compound fracture of the parietal bone,	1½	Death.
4	Do. do. ...	11	Recovery.
5	Simple fracture of the frontal and ethmoid bones, ...	45	Death.
6	Fracture of the base of the skull,	33	Death.
7	Probable fracture of the base of the skull, . .	39	Recovery.

Cerebral concussion, with or without scalp wounds—5 cases—all of which recovered.

Case 1—Compound Fracture of the Parietal and Temporal Bones.—
A lad, aged 16, was accidently thrown out of one of the swing boats, belonging to a merry-go-round, upon his head. A serious scalp wound was inflicted on the back of the head extending nearly from one ear to the other, and lifting up a large flap of scalp· Nevertheless so little was the lad affected by his injury that on the same evening and for several successive days he walked a considerable distance to a suburban hospital to have it dressed. He began to have sleepless nights, however, and became very weak. On the fifth day his sight was affected, and on the seventh he was brought to the Infirmary very ill indeed. The extensive scalp wound was in a foul state, and bare bone was visible beneath it. On feeling under the flap with the finger a fracture was made out just behind the middle of the right parietal, but no depression could be felt. He rapidly became delirous, with a dry tongue and sordes on the teeth. The pupils remained equal and sensitive, and there was no distinct paralysis of the limbs, but he passed his water in bed. He was unceasingly restless and had a short hacking cough, which an examination of the chest shewed to be due to intense congestion of the bases of the lungs. On the fifth day after admission he was con-

vulsed on the right side and died on the sixth.   Twitchings of the right side of the body persisted till death, and for some hours before that took place he vomited coffee-ground stuff incessantly. At the autopsy a fissure was discovered commencing at the vertex, running outwards across the right parietal bone and downwards through the squamous portion of the right temporal almost to the jugular foramen where it ended in the masto-occipital suture.   All the sutures in the immediate neighbourhood of the fracture were loosened and slightly apart.   On removing the skull cap, about half an ounce of pus was spread over the dura mater beneath the right parietal bone.   There was a rent in the dura mater and, on lifting it up, the superior surface of the right cerebral hemisphere was found covered with pus over an area of several square inches, while just in front of the interparietal fissure (on the right side) the brain was softened and broken down to the depth of about half an inch.   The cerebral tissue surrounding the broken down portion was dark coloured and congested.   The membranes around the base of the brain and over the cerebellum had patches of lymph upon them here and there.   We were very much struck with the fact that with so serious an injury the boy should, for five days, have been able to walk to a hospital for treatment.   Had he been kept perfectly still from the commencement things might have been different with him.   The question of trephining was frequently discussed, but as only a fissured fracture could be made out (and that with difficulty) while no depression was present, the idea was abandoned.   The autopsy shewed that it would not have done any good.   It was also noted that while all the coarse disease was on the right side, it was also upon that side of the body that the con-vulsions and twitchings which preceded death occurred.   The lungs were found intensely congested and friable.

Case 2.—An old man, aged 75, was struck on the head by a steam engine.   There were two or three scalp wounds, one of which led to a depressed fracture of the right parietal bone a little way above its anterior inferior angle.   On admission his skin was warm and moist, his temperature was normal and he had a full pulse of 84. His left lower limb was quite rigid and his pupils were equal, but contracted and insensible to light.   His respirations were stertorous and only 12 in the minute.   In the course of three hours they sank to the alarmingly slow rate of 5 in the minute, so trephining was at once performed and the depressed fragment raised, with the result that they almost immediately rose to 12 and the pulse to 120. For six days the patient remained, with one short interval, unconscious and passing his evacuations in bed, and then died, the pulse ranging from 100 to 120, and the respirations from 18 to 24. On the surface of the dura mater opposite the trephine hole a little healthy pus was found, but the membrane was not inflamed at all. Beneath it, however, was a diffused suppuration in the arachnoid cavity. The brain was not lacerated nor apparently softened. The most interesting feature was undoubtedly the excessive slowness of the respiration shortly after the patient's admission, and its acceleration immediately after the removal of the depressed piece of bone.

Case 3. Was in many respects a very unfortunate one. The patient, an infant eighteen months old, was seated on a stool by the fireside when its mother, in reaching some article from a shelf above the fire, dropped upon its head a sharp pointed poker, which produced a compound, depressed fracture of the left parietal bone over the parietal eminence. The depressed portion of bone was very small—about the area of a threepenny piece,—the depression was only slight, and the wound on the scalp comparatively trifling. On the fifth day after the accident the child lost the power of the right leg, and on the following day that of the right arm to a great extent. A fortnight after the injury mother and child both came into the Infirmary. The child seemed fairly well except for the paralysis of the arm and leg, and even that seemed motor only, as sensation did not seem much, if at all, impaired. The question was, Should we trephine? As there was nothing urgent I determined to wait for a day or so. But, as each day passed by, the child looked better and by the end of a week began to make some use of the affected hand. It remained for eighteen days under observation in the Infirmary, and at the end of that time it seemed well and full of play, able to walk about with the assistance of its mother's hand and able to raise the affected arm above the head. The small wound discharged a little healthy pus and was dressed with water dressing, but the probe passed into it impinged upon bare bone. With such a process of rapid spontaneous recovery going on, it seemed as if trephining might only bring about some dangerous condition instead of averting one. As the mother was anxious to get back to her home in the country, she accordingly was allowed to leave with the child, although my desire was that it should remain a little longer with us under observation. A few days after its return the medical gentleman, who sent the patient to the Infirmary, saw it and found that the paralysis had quite disappeared, the child running about as well as ever. A week after that it had a fit, lasting only a short time, which was thought to be possibly caused by the eruption of some molar teeth. Eight days later it was taken with another fit;—a very severe one, in which it died. The doctor's account of the autopsy, says—" On removing the "scab which had formed over the little scalp wound a healthy granu-"lation was found beneath. The calvarium being taken off, an oval "aperture was found not far from the upper margin of the left "parietal bone, to the margins of which the dura mater was very "adherent. As soon as that membrane was detached, about four "ounces of foetid pus escaped, and a hole an inch deep was found "in the substance of the brain. The whole cerebral surface, over "an area of two or three inches, was pulpy and mixed up with pus."

The case illustrates the extremely treacherous character of all head injuries in children. At the time the infant was in the Infirmary, there was nothing to call for trephining, and I anticipated that the wound would heal, probably after a scale of the outer table had necrosed and come away. Even if I had trephined then, I should probably have found healthy dura mater, and I do not think under the circumstances, that I would have cut it open. The mischief seems to have been purely cerebral, and not directly proceeding from the injury to the bone. It is astonishing how the child could have gone on with so much mischief present; but it only shows how careful one must be to maintain a watch over such a case for a long time after the first grave symptoms have disappeared, and not to pronounce too decidedly about the future.

Case 6.—Fracture of the Base—Bleeding from the Ear—Death.— The patient, a man, aged 33, was knocked down by a hansom, and his head came violently in contact with the kerb-stone. He was brought in insensible, and remained so until he died, forty-four hours afterwards. All the external marks of injury were on the left side of the head, and he had sanguineous oozing for a while from the left ear. At the autopsy a most extensive fracture was found, commencing behind at the back part of the left parietal bone, and running forward. It then bifurcated, one limb running upwards towards the vertex, and the other downwards towards the base, thus taking a sort of circular direction around the place at which the patient's head was believed to have come in contact with the pavement. Now, although the external injuries and the fracture were on the left side, it was found that the brain injuries were on the opposite, the temporo-sphenoidal and inferior frontal convolutions of the right side being reduced to a mere bloody pulp. Beneath the dura mater a large amount of blood had been poured out, and it was noticed that it was by no means most abundant opposite the site of the fracture. This showed the futility of any attempts at trephining, unless there is something in the way of depressed bone distinctly to be made out. Unless this is the case, trephining must be an absolutely haphazard proceeding, as no one can with certainty say where the internal lesion is. It was noticed that, although there had been some sanguineous discharge from the ear, the main fracture did not extend to the petrous bone. This was cut out, and I very carefully examined it at my leisure. On stripping the dura mater from the anterior surface of the bone, a minute, but distinct, fissure was detected running parallel with the long axis of the bone. A section of the tympanic chamber being made with a fine saw, this fissure was found to lead into it, and the chamber itself was occupied by a small blood clot. The membrane was then carefully examined, but no distinct rupture could be made out, so that it was difficult to see how the blood escaped into the meatus. The only explanation I can offer is that the membrane was only slightly split, and that, as soon as the blood ceased to exude through this, its edges fell together and presented no trace of separation. It was interesting to note how the hard but brittle temporal bone had a crack in it quite special to itself, and distinct from the main fracture. Such a condition may be present in some of those cases of so-called fracture of the base, diagnosed by bleeding from the ear, in which recovery takes place. Such a one was case 7, where a groom fell from his horse, and was brought to the Infirmary in a semi-conscious state, bleeding profusely from the right ear. This bleeding became less and less, and ceased forty-eight hours after the injury. The only other serious symptom was severe headache. After ten days the patient left the Infirmary, but for eight or nine months was far from well. For instance, his speech was hesitating; he had frequent headache and "queer" feelings in his head; his smell was distinctly affected, and his power of tasting diminished ; he was giddy when he stooped; he had flushing of the face after his meals; and a constant singing and deafness in his right ear. It was nearly a year before all these completely disappeared. What was the nature of *his* injury? Was it a complete fracture of the base, or only an isolated crack in the temporal bone, such as that described above? And would a railway company have believed in his subsequent symptoms if he had been injured on their line and demanded compensation?

It may not be out of place to mention a small practical point in connection with the treatment of suppurating cervical glands, which, although practised doubtless by the majority of hospital surgeons, is not much mentioned in books, and not very generally known among men in private practice. It is with reference to the use of the drainage tube. There is hardly any neck gland abscess which cannot be thoroughly emptied by a puncture that may be made with a tenotomy knife. There has, of course, been a great tendency to wait until the matter should come well to the surface. By that time the skin is reddened and thinned. A free incision being then made, the skin retracts and shrivels up, and an unsightly cicatrix ultimately results. I would strongly recommend an opening to be made so soon as fluctuation is certainly recognised. The child should be invariably anæsthetized, so that a minute puncture may be steadily and quietly made in the abscess, through which a small drainage tube should be inserted and retained in position.

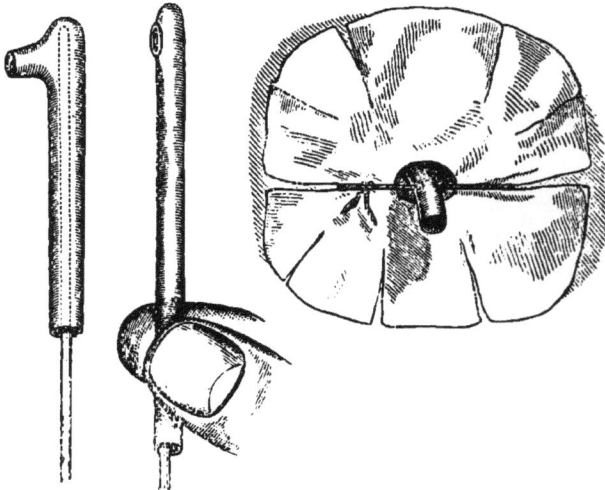

To insert a tube, pass a probe through it, to within a quarter of an inch from one end, and let it impinge there against the wall of the

tube, pushing this before it.   Then cut off the small end close up
to the probe, and pull up the tube on the probe, holding it firmly between the finger and thumb.   Having inserted the tube in the wound
at the proper depth, pass right across it a piece of pack-thread by
means of a small darning needle.   Cut a circle of adhesive plaster, and
nick it round the edges to make it lie comfortably.   Then cut a
hole in the centre and a slit at each side.   Lay it down so that the
tube protrudes through the hole, and bring up the thread on each side
through the slits, and tie it front.   Cut the tube short.   A little
absorbent cotton wadding, wrung out of a disinfectant solution,
should be put on the surface and frequently changed.   Thus a vent
is given for the discharge, which is sucked up by the wadding, and
the abscess cavity can be washed out, while the parts are not kept
soddened and heated by poultices or waterproof dressings.   Every
two or three days the plaster is removed, and the tube replaced
and shortened, if necessary.   When all is closed only a white spot
marks where the tube once was inserted.   I was shown this little
manœuvre by Mr. Bickersteth many years ago, and have found it
most valuable in private practice.   I often show it to students, and
tell them it will be far more useful to them than lucubrations on
tying the innominate or performing gastrostomy.

ABSCESSES.

Axillary abscess, ...2 cases.
Abscess among forearm muscles, ...1 case.
Facial abscess from bad teeth, ...2 cases.
Abscess of the tongue,..1 case.
Superficial abdominal, perineal, mammary, and subpectoral abscesses, ...5 cases.
Abscess of the neck—secondary hæmorrhage, ...1 case.
Submaxillary cellulitis, ...2 cases.
Whitlow—severe, ...4 ,,

The only case of interest among the abscesses was one of
hæmorrhage from an abscess in the neck.   The patient was a
weakly man, aged 32.   Three weeks before admission there came
a small lump below the right ear, which gave him a good deal of
pain, and gradually increased in a downward and backward direction.   It softened and suppurated, and was opened by his medical
attendant, a considerable quantity of pus being evacuated.   He
applied poultices, and after a week, during a violent fit of coughing, blood burst from the wound, in his own words, "like a tap
being turned on."   Two days afterwards there was a second attack

L

while coughing; again, a third attack after another two days' interval; and a fourth one on the morning of admission. Pressure and other remedies were tried in vain. When brought into the ward he was in the last stage of exhaustion, barely able to speak, and quite unable to stand. The right side of the neck from the ear to the clavicle was occupied by a great fluctuating swelling. In front of the sterno-mastoid, about half way down, was the original incision, from which a little sanious, bloody discharge was issuing. Behind the muscle, in the posterior triangle, a piece of skin about an inch square, was actually sloughing from the subjacent pressure. A glance at the man made it clear that another attack of hæmorrhage would infallibly finish him, and that the bleeding vessel must be found and tied at once, and at all costs. He was so weak that it was not deemed advisable to take him into the theatre, but his bed was simply pushed before a window through which there came a brilliant light. Under ether the original incision was enlarged upwards to the mastoid process, and downwards for about two inches. A great quantity of putrid, broken-down blood-clot, mingled with pus, was turned out. Then a similar incision was made behind the sterno-mastoid, through the sloughing skin From ear to clavicle, and from the nape of the neck almost to the trochea, the tissues had been torn up by the pressure of blood and utterly disorganized. Everything being mopped and cleaned up, blood was found to be trickling down from somewhere very high up. To get at it the sterno-mastoid and skin over it were cut clean across, thus uniting the two vertical incisions by a transverse one. The muscle was dissected upwards, exposing the sheath of the carotid vessels in nearly their whole extent, but still the blood always kept running from some deep-seated point high up. At last this was reached, just in front of the transverse process of the atlas. From it arterial blood issued, and an aneurism needle was thrust through the tissues on each side of it and ligatures applied, which at once checked all further bleeding. The vessel was the occipital artery, not far from its origin from the external carotid. Into it the abscess had made its way. The great wound was rapidly swabbed out with turpentine, and then stuffed with lint dipped in the same fluid. By the time the operation was finished, the patient's pulse could barely be detected, and his face was perfectly blanched and covered with a cold sweat. I expected to hear that he had died in a few hours afterwards, but next morning found that he had passed a good night. He made a rapid recovery, and in a month left the hospital with the wound quite healed. The case was a most instructive one, as demonstrating the value of energetic and thorough measures in a case where the vital powers were so reduced by loss of blood that it seemed impossible that any prolonged operation could have a chance of success. And, indeed, I was quite prepared for the patient dying under it, but was determined that so long as he had any blood to run out of him the place whence it came should be found and tied.

In connection with this case I would like to say a few words in favour of the old-fashioned remedy turpentine, as a cleansing styptic. In former days it was the regular thing for oozing, until superseded by the introduction of perchloride of iron. This has always seemed to me most unfortunate, as the iron is the very worst of all styptics. Owing to its great potency, and the rapidity with which it acts, it soon became popular, and is at the present moment the favourite stand-by of the chemist, who diligently swabs with it every cut that is brought into his shop, and then stuffs the wound with lint soaked in it, preparatory to sending the patient to a hospital. As a result, the wound is covered with a cake of coagulated blood, and its surfaces are sometimes positively killed by the strength of the application. Beneath this firmly adherent crust all sorts of purulent, filthy secretions accumulate, till at the end of forty-eight hours it stinks abominably, and requires to be well poulticed to get it clean. Should bleeding recur, the difficulty of finding the spot is enormously increased by the mess of pus and almost cineritious hard clots which cover it. I have seen so many cut hands almost ruined by it that I have totally abandoned it. On the other hand, turpentine is nearly as powerful a styptic, and is a most marvellous cleanser and sweetener. The plug soaked in turpentine comes out quite easily at the end of four-and-twenty hours, leaving a wholesome granulating surface behind it. For all wounds about the perineum, such as lithotomy wounds, fistula, cuts, or incisions for extravasation of urine, there is nothing like it, and I trust it will soon be reinstated in surgical favour. Our forefathers had some excellent remedies, and this is one of them.

ULCERS.

Ulcers of the lower extremity, ordinary, strumous, syphilitic, varicose, &c.,
of which one proved fatal from extensive sloughing, ...16 cases.
Ulcers of the trunk and upper extremity,................................. 3 ,,

With regard to the treatment of the nineteen cases of ulcer, it
cannot be said that there is anything of note to communicate, except to congratulate oneself upon the immense improvements in
managing them afforded by Martin's rubber bandage, skin-grafting,
sponge-grafting, iodoform, strapping, and so forth, which have made
it possible to heal up surfaces which our immediate predecessors
would have regarded as quite hopeless. For instance, one evening
a carter, aged 34, was admitted with a huge ulcer of the leg, which
had eaten into one of the superficial veins, from which he had lost
a very considerable quantity of blood. The bleeding was easily
stopped, and the dresser's notes describe the ulcer as " occupying
" the greater part of the left leg, reaching from the ankle to three
" inches below the knee-joint, and completely encircling the limb
" at one part." It was in a shockingly sloughy state, and positively
the lower part of the gastrocnemius and tendo Achillis were nearly
eaten away, while in front a large piece of tibia was bare. The
patient was strongly urged to have amputation performed, but he
would not hear of it, so we set to work to clean up this great putrid
surface with charcoal poultices, feeling sure that the man would
soon find out what a mistake he had made. To our surprise, the
sloughy surface rapidly cleaned up, and became so ruddy and
wholesome that a wholesale skin-grafting was done upon it. Many
grafts took, and soon little islands of skin appeared, which spread
over the whole area, so that at the end of three months it was perfectly healed over. Even when he left, dismal prognostications
were uttered that the cicatrix would not stand hard work, but would
soon break down, leaving him as bad as ever. However, not long
ago he turned up to get some advice about another ailment, more
than a year after his discharge, with the leg quite sound. It looked
more like a stick than anything else, it is true, but he said it was
as useful as ever. Without grafting I do not think this could ever
have been cured. By the way, referring to the bleeding in this
case, what effect has a good sharp bleeding upon sloughing? There
is no doubt that many unwholesome wounds and stumps begin to
look amazingly better after an attack of secondary hæmorrhage.

But then one must remember that to arrest this secondary hæmorrhage it is generally necessary to lay bare all the unwholesome surface, scrub it, cleanse it, and restore it again to a tolerably healthy state. So that it may not be the loss of blood after all, but the purification of the wound necessitated by it, which does the good.

Concerning sponge-grafting I have given it a fair trial; but, on the whole, I cannot see that it possesses any advantages over skin-grafting, while the latter is an infinitely simpler process. Whatever is to be of universal use must be simple, and the country surgeon can always do skin-grafting and always has the material at hand, whereas, sponge-grafting is a decidedly elaborate performance. It is a most interesting performance no doubt, but I do not think it will stand the test of time. A thorough blistering to a chronic ulcer, followed by skin-grafting, will work a more speedy and certain cure.

As regards the chronic ulcer with the cartilaginous edges, I find that students are acquainted with all sorts of novelties in the way of applications for its treatment, but very few seem to know that a big fly blister is the best of all.

One modern plan of treatment has impressed me most favourably, viz.—the treatment of lupus and lupoid ulcerations by the vigorous application of the actual cautery. One of the cases in the present list was that of a girl of 19, who had had a lupoid ulceration in front of the wrist for two years, upon which every kind of remedy had been expended. She was put under ether, and the whole affected surface destroyed with the thermo-cautery used at a white heat. In a few weeks it was perfectly well. The other day a small boy of nine came under my care, with one-third of his nose gone from lupus, and a sore commencing to eat into his upper lip. One smart application of the cautery stopped all further progress of the disease, which straightway healed. It is a pity this treatment is not more commonly used. Its apparent barbarity is much against it, but this is only apparent and not real. Of its efficacy in arresting all further rodent action, and reducing an intractable ulceration to a healing surface, I am convinced by the experience of a good many cases; and, furthermore, no more deformity or cicatrix is left than if the disease had healed up under some of the usual ointments or lotions, for the tissues that are destroyed by the cautery are not healthy or permanent tissues, but diseased granulations.

DISEASES OF THE BURSÆ.

The patellar bursa is subject to every form of enlargement from slight distension with serous fluid up to the formation of a solid tumour. Pressure from kneeling being the almost invariable cause of this, domestic servants are, in consequence, its usual victims. Most bursal swellings are, fortunately, mere effusions of serum into the sac, and relief from pressure is practically all that is needed for their cure. One must take care, however, not merely to forbid kneeling, but all and every source of pressure whatever. In the days of crinoline, when very heavy affairs stiffened with metal hoops were worn by the poorer classes, in imitation of the lighter and more costly apparatus of the rich, I remember being considerably troubled by servant girls, whose bursæ would by no means disappear even after kneeling had been quite given up. Usually the patient was found to wear a cheap, heavy crinoline, which kept bumping against the knee at every step, and which had to be discarded before any good could be done.

If the bursa remains in an irritated condition for any length of time the sac becomes thickened, and some active treatment is necessitated. Probably the best plan is to tap it with a small trocar, and apply a blister over it, keeping the leg at rest on a back splint. As soon as the blistered surface will bear it, the knee should be strapped so as to induce the walls of the sac to cohere. Should this proceeding fail, and the fluid reaccumulate, the bursa may be again tapped, and a teaspoonful of strong iodine tincture thrown in, just as one does with a hydrocele. Some surgeons have warmly advocated the introduction of thread or worsted setons, but I cannot say that this is a proceeding of which I am much enamoured. My objection to it is that you set up suppuration in the bursa, and at the same time do not provide a sufficient drain, so that a good deal more inflammatory mischief may be aroused than is required for cure. The subjects of the complaint under consideration are almost always treated as out-patients at hospitals. A thread seton is put in on a Monday, let us say, and the patient is directed to return on Wednesday. But by Wednesday the knee is very painful, and she thinks she will wait till Thursday before coming up. On Thursday the pain is so bad that she cannot walk at all, and on Friday she is brought to hospital in a cab, with the knee frightfully red, swollen, and tense. Then she is taken in, the suppurating sac is freely laid open, and she is a month in bed. Having seen this actually occur, I am somewhat chary about using setons.

But if the tapping and injecting, the blistering and strapping all fail, what is to be done? My present practice is to *excise the bursa* at once, and I would strongly urge the more frequent adoption of this proceeding. In performing an operation which is not necessary for the saving of life (an "operation of complaisance," as the old surgeons would have termed it), one has to balance against the annoyance produced by the complaint, the pain of the operation at the moment of doing it, and the subsequent risk caused by it. With regard to pain, that is a thing of the past, as far as the work of the knife is concerned. As for the subsequent risk, antiseptics have put such an operation as the removal of the bursa patellar almost on a level with the commoner surgical proceedings of paring one's corns and cutting one's nails—operations, by the way, which have both been followed by fatal results, but which, in spite of that, are universally practised.

Excision of the bursa is, of course, no new thing. It has long been performed, and is doubtless recommended in all surgical text-books, but as a rule the recommendation is not pressing. On the contrary, the praise is decidedly faint. Holding the opinion which I have just expressed, I should naturally like to see excision advocated, not in the light of a dire extremity, but of a thing to be done on comparatively slight provocation. Hitherto it has practically been reserved for solid bursæ of considerable size, but I venture to think that it may be employed with advantage before the condition of solid tumour is arrived at. All bursal swellings are primarily cystic. They become solid, as I have often demonstrated in the theatre after an excision, in two ways, first by repeated deposits of lymph on the interior of the sac, and secondly by threads of lymph stretching across from wall to wall, which form tuberculæ and cut them up into cells, which in turn become filled with solid matter. My idea is to have them out before they have become actually solid—when they are, so to speak, semi-solid. For, after all, when they have reached this stage, tapping and injecting are thrown away upon them, and the only thing that will obliterate them is some such process as suppuration induced by a seton. Now, if the excision of a thick-walled or semi-solid bursa can be effected painlessly and safely, the next point to find out is whether it can compete with the seton in point of time. I think it can, and have come to the conclusion that a certain and permanent cure by removal does not take longer than that by prolonged suppuration, and is in all other respects much to be preferred.

In the method of operating I have ventured to make an innovation upon the ordinary practice of removal by a median incision. I have for some years made two incisions, one on each side of the joint midway between the edge of the patella and the femoral condyle. There seem to be two advantages which this method possesses. In place of there being after the operation a median cicatrix, which is painful to kneel upon, the skin in front of the patella is as sound as ever. This is no small advantage, seeing that the persons who get these bursal tumours generally make their living by kneeling, and have to take to kneeling again, even after they are removed. Indeed, I was first induced to try the double lateral incision by the case of a coloured man, the cook and steward of a ship, who, in the performance of his multifarious duties, was necessarily a good deal on his knees, and so grew an immense bursal tumour. This was removed by the usual median incision. A prolonged suppuration ensued, the lips of the wound gaped, and a thick cicatrix resulted. He came to see me some months afterwards, and, although very grateful for the removal of his lump, complained not a little of the pain and inconvenience which the cicatrix occasioned him. In my next case the double lateral incision was employed with such satisfactory results that I have continued to employ it ever since.

The second advantage given by this method of removal is the excellent drainage obtained by it. This can never be got by the median incision, inasmuch as any effused fluid at once sinks down into the dependant parts of the pouch and pockets there. If suppuration occurs, the greatest trouble is sometimes experienced in getting quit of the pus. Morning and evening the cavity is washed out, and morning and evening it is found full again, until after considerable delay incisions have to be made at the sides, which might as well have been made at the beginning, So much have I been struck with the great advantage of drainage by lateral incisions that *I have applied the principle to all suppurating patellar bursæ.* In such cases a median incision (more especially if it be, what it too frequently is, a mere prick with a lancet) is most inefficient, and causes a great deal of unnecessary delay and trouble in the cure. Having ascertained with certainty the presence of pus, the patient should be anæsthetized and the suppurating sac cut down upon at as low a level as possible on one side. The finger should be inserted into the abscess cavity, and pushed over the patellæ to the most dependant point on the other side. The point of the finger should then be cut down upon. The incisions being kept open by a couple of large drainage tubes, the process of coalescence of the walls of the abscess is wonderfully accelerated.

While removing a bursa by the double incision, the great point is to keep steadily cutting upon the tumour itself. In front of the patella no possible mischief can be done, but just at its edge the capsule of the knee-joint is very thin indeed, and here there is a strong tendency to get off the tumour and go cutting down through the aponeurotic capsule. Although quite aware of this risk, I once inadvertently made a small hole into the joint. Fortunately, I was operating under the spray, and by taking care that no blood got in, no harm resulted.

The two following cases are examples of the periods that may reasonably be given for a cure, either when suppuration has occurred or when union by first intention has been secured:—

Mary D., a healthy-looking girl of 19, was admitted into the Royal Infirmary with a hard bursal tumour over the right knee about the size of a small orange. She was a general servant, and accustomed to a great deal of kneeling. About twelve months previously, after an extra amount of house cleaning, she first observed a slight swelling over the right patella, which steadily increased in size and hardness. It was painless, but very inconvenient to her when at work. The only treatment had consisted of painting the skin over the swelling with iodine tincture. The tumour was removed under antiseptics by the double lateral incision, but no drainage was employed, as it was anticipated that complete primary union would follow. This was a mistake. The patient proved extremely restless, always kicking the leg about, and on the third day suppuration ensued and materially delayed the cure, inasmuch as it necessitated the removal of the stitches, and so the lateral incisions gaped. By the eighth day discharge had ceased, and the incisions were simply two elongated sores, which were dressed with red lotion. On the twentieth day the wounds were completely healed and the patient discharged.

As an example of rapid healing the case of Jane L., aged 26, may be cited. Three years before admission, she had suffered from a bursal swelling over the left knee, which had been cured at St. George's Hospital by rest, tight bandaging, and a back splint. Quite recently a similar swelling had formed over the right knee, and, from its hardness, it was resolved that time should not be wasted on preliminary measures, but that it should be removed at once. This was done exactly after the same manner as in the previous case, except that two small drain tubes were inserted. On the second day the antiseptic dressing was renewed, when the wound was found to be healing by primary union, and the stitches were removed. The dressing was not renewed till the seventh day, and then the whole thing was seen to be practically sound. The antiseptics were left off, and some boracic ointment applied, while a couple of straps were put round the knee to protect the still tender incisions. On the tenth day the patient left the Infirmary fit for work. Two dressings and not a drop of pus.

These two cases fairly represent what may be expected after removal of a bursal tumour by double incision and listerism. At the outside, three weeks to a month is required, and, if the treatment be very successful, only ten days. Let us even take an average of three weeks. I venture to say that if a bursa has got to that state that its removal is justifiable at all, excision will prove a cure as quick and infinitely more certain than any other form of treatment with which I am acquainted. I would, therefore, urge upon the operating surgeon, (1) the earlier removal of bursal swellings than has hitherto been practised; (2) their removal by double lateral incision under antiseptic precautions; (3) the application of the double lateral incision to all suppurating patellar bursæ as an effectual means of drainage, and so of accelerating recovery.

Of the two cases of injury to the olecranon bursæ causing suppuration, one, being promptly and freely incised, got well soon; but the other, arriving at the Infirmary after great damage to the parts around the joint had occurred, sustained an extension of the mischief to the articulation itself, which ended in its destruction.

Thomas G., aged 41, a striker, in December, 1880, while at his work, received a blow from a bar of iron upon the right elbow. A good deal of swelling ensued, and he poulticed it with bran and vinegar. Nevertheless, he continued at his work till January 22, 1881, when, as he was screwing up a frame, the screw broke and the frame rebounded, striking him on the arm, a little below the site of the previous injury. For two months after this he was under the care of the surgeon of his club, keeping the elbow at rest in a sling and poulticing it. The bursa, which had evidently suppurated, then burst into the neighbouring tissues, and incisions were made on four occasions. Nevertheless, the man's surroundings were such as to make efficient surgical treatment very difficult, and so he was put under my care by his medical attendant on March 22nd. The elbow and neighbouring parts were much swollen, and of a livid hue, brawny to the touch in some places, and boggy in others; evidently rotten and undermined. Under ether three long slices were made through the tissues, and sanious fluid oozed out. Lead and opium lotion was applied, and the arm kept at rest on a pillow. In spite of this, the patient's temperature kept high, and he suffered a great deal of pain, and very shortly it became evident that the elbow-joint itself was affected. The limb became utterly helpless, as limbs do when a great joint is attacked, and grating, indicative of destruction of the cartilaginous surfaces, was detected. The arm was then firmly fixed upon a Thomas's elbow-splint, and rendered completely immovable, while the wounds were dressed with various lotions. By degrees the swelling of the tissues went down, and the various sinuses and incisions closed up, all but one, which was situated right over the olecranon. Here a probe passed down to the bone. Accordingly, in the end of May an incision was made, and the greater part of the olecranon being found bare and dead, was clipped away with bone forceps. As it was clear that a stiff joint must ensue, this did no harm, but the reverse, inasmuch as it gave free vent to any discharge which might tend to accumulate in the articulation. The wound was packed, and made to heal up from the bottom.

The case just narrated shows what a serious thing an inflamed olecranon bursa may turn out to be, and I remember a case of a very similar nature occurring in a drunken carter, where the joint was utterly destroyed, pyæmia set in, and the man died. On two or three occasions I have seen the joint barely escape, while the tissues all round it were so riddled with sinuses as to render very severe incision necessary, and many weeks ensued before the joint could be used. In some instances, of course, the patients are, by their neglect, entirely to blame for such a condition, but there are others where it is equally due to a want of prompt and vigorous action upon the part of the surgeon, who, to save his patient pain, contents himself with making a mere puncture, instead of a prolonged incision. Whenever I find a drunken-looking fellow in the waiting-room, who has fallen on the kerb-stone a few days previously, and who has a thing like a very red Tangerine orange over the back of his elbow, I have him taken into the ward, give him ether, and lay the abscess open from end to end. The usual "pricking" and "lancing" suitable for boils and small abscesses, is of no use here, if it is desired to save the patient with certainty from the risks of a cellulitis. In the striker's case, a period of more than a year and a half elapsed before he could swing a hammer. The bread-winner of a family kept out of work for a year and a half means a great deal of suffering, and a serious inroad upon the public charitable purse. Hence, I have always made it a point in clinical teaching to impress upon students the early recognition of suppuration in the olecranon bursa, its danger when it does occur, and its one method of treatment—viz., ether and incision from end to end. *

* I have embodied the above views in a little paper in the "Liverpool Medico-Chirurgical Reports for January, 1882."

Injury.	Cases.	Recovered.	Died.
Crushes, { Crushed hand, ...	3	3	0
Do. foot, ...	3	3	0
Do. toes, ...	3	3	0
Cut throat,	5	3	2
Burns and scalds, ...	7	7	0
Scalp wounds, complicated with other injuries and with delirium tremens, ...	11	9	2
Tetanus, ...	2	0	2
Incised, lacerated and punctured wounds,	18	18	0
Abdominal injuries from crushing, ...	3	1	2
Gunshot wounds,	2	2	0

THE cases enumerated under the head of "wounds" are in Liverpool so numerous that, unless very severe, they are not admitted into the Infirmary. So that it may be taken for granted that, although the above list contains only a few instances which possess anything out of the common about them, they were all sufficiently grave to require prolonged attention and sometimes give rise to no little anxiety. Among the punctured wounds for instance was a penetrating wound of the chest and a stab wound of the abdomen. In the latter case the intestine protruded through the wound and yet was not injured—as not infrequently happens.

Severe Case of Cut-Throat.—The patient was a poor, wretched looking widow, 54 years of age, a servant of all work. She had strong religious delusions of a melancholy turn, and one night in the kitchen cut her throat with a table knife. She was found on the floor next morning apparently dead in a pool of blood, and was brought to the Infirmary about 10 o'clock. The gash was a very deep one. It had nicked the sterno-mastoid, divided the sterno-hyoid and sterno-thyroid, and cut the lower part of the thyroid cartilage clean through from front to back, severing the arytenoids. The patient was so completely collapsed that one could not venture to do anything with her, otherwise I would have performed tracheotomy, put in a tube and stitched up the transverse suicidal wound so as to have induced it to unite at once. Her death was expected every minute. To our surprise she lived through the day, and the next and the next, and eventually began to mend. But it was then too late to do any good by tracheotomy. The wound rapidly granulated over, but the edges gaped widely apart, and thus it came to pass that the cut ends of the air tube, in place of facing each other upwards and downwards, began to turn out in a forward direction. Very soon, as the tissues drew together in the process of cicatrization, the lower orifice came to the front and the upper one was tucked up almost out of sight. Moreover, they began to contract amazingly, particularly the upper one, so that it was clear that it would soon close up. Therefore, it became necessary to do something to maintain the ends of the air tube patent and to keep them in the same line. A rubber tube was accordingly inserted into each of them, and this was gradually increased in size until their normal calibre was regained. After that, a single rubber pipe was employed, one end of which was pushed down the air tube for about an inch and the other end upwards for about three-quarters of an inch. A string was attached to it in front. Then the tissues were carefully dressed so as to induce them and the air tube to close in around the rubber pipe, which at once preserved the bore of the tube, and kept its severed portions in the same straight axis. By degrees the parts gathered in, leaving only a small hole out of which the string hung, and we were beginning to wonder how the rubber pipe was to be got out, when the unfortunate woman caught cold on the hospital green and, being still very feeble, succumbed to a sharp attack of bronchitis in a couple of days—to our great disappointment. The case is an example of what is so often seen, viz., complete division of the air tube with escape of the great vessels, for which condition tracheotomy seems the best plan of treatment, bringing the edges of the divided air tube together, and closing the suicidal wound completely. By this means the difficulties experienced in this case, owing to the contraction and want of continuity between the ends of the tubes, would have been avoided. The rubber pipe, acting in the same way as in the case of the artificial anus described at p. 95, would undoubtedly have succeeded had not the old lady suddenly fallen a victim to the bronchitic attack.

Rupture of the Duodenum, and Rupture of a Renal Cyst.— A powerful young man was admitted in a state of complete collapse, and died a few hours after his admission. While working in an excavation a great mass of earth had fallen upon him, and jammed his abdomen against an iron rail. In his waistcoat pocket the bowl of a thick, short, clay pipe was found broken into very small fragments—a proof of the severity of the crushing. Yet there was not the least mark nor sign of bruising upon the skin of the belly. After death the omentum between the stomach and the transverse colon was found to be extremely torn, while the colon itself was bruised. The duodenum was ruptured in a transverse direction, the tear extending across about two-thirds of its circumference. The main trunk of the superior mesenteric vein was nearly torn across, and the inner coat of the aorta displayed a long rent. Rupture of the duodenum must be so rare an accident that it was thought worth while to record this case, as showing the mode in which the injury may be produced.

A railway guard was crushed between the buffers of two carriages, and was admitted in a state of profound shock, from which he never rallied, but died in about ten hours. After death the abdomen was found full of a sero-sanguineous fluid. It was discovered that the right kidney had been converted into a great cyst, with calcareous deposit in its walls, and that this cyst by evil chance had been ruptured, and so its contents had been poured into the peritoneal cavity. The left kidney, which was sound, weighed eight ounces. We heard from the patient's wife that he had enjoyed good health up to the date of the accident, and had not suffered from any symptoms of kidney disease.

Injury to the Knee from a Rocket Tube.—William Rogers, aged 33, an engineer on board an Indian merchant steamer, was standing on deck watching some rockets being fired in celebration of New Year's Day, 1881, when the mortar burst. Something struck him on the outer side of the right knee, and inflicted a small wound. He was carried below, and a surgeon from a Queen's ship that was close at hand extracted from the wound a small piece of wood. He was landed at Suez and taken to hospital, where he remained for six weeks, having the knee painted with iodine and rubbed with ointment. At the end of that time he was brought home to Liverpool, and was admitted into the Royal Infirmary about two months after the accident. The right knee-joint was hot, inflamed, much swollen, and the source of great pain. The swelling was not due to fluid in the interior of the joint, but to a general thickening of the femur and of the tissues around the articulation. Over the outer condyle was seen the orifice of a sinus, down which a probe passed for about two inches to the back of the joint. The limb was fixed on a Thomas's splint, and lead and opium lotion applied to the knee, which gave the patient immense relief from his pain, and considerably reduced the swelling, so that after about ten days rest the sinus orifice was enlarged, and the finger passed along it. It was found to lead to a cavity immediately behind the posterior ligament of the joint, which contained some thin pus; but the finger did not encounter any bare bone or foreign body. Care being taken not to come in the way of the great vessels, a counter opening was made into the cavity at the back of the popliteal space, and it was drained by a large tube. The limb being once more put upon the splint, was kept at rest for ten weeks, and then another careful examination was made. Although comparatively free from pain, and much smaller than it was at first, the joint was still considerably swollen, while from the drain-holes came a free discharge of pus. A probe was passed into the sinus over the outer condyle, and distinctly touched bare bone. Ether was therefore administered, and the sinus was laid freely open, so as to expose the back part of the external condyle of the femur. From this I removed some small loose fragments of dead bone, and then pulled out what I imagined to be a large sequestrum. It felt, however, remarkably heavy, and to our surprise turned out to be a piece of gun-metal, about three-fourths of an inch square by one-third of an inch thick, which had evidently been embedded in the substance of the condyle. The further progress of the case need not be specially detailed. The wounds closed up, and the joint all but resumed its normal appearance, although very stiff. Before the patient left the Infirmary it was forcibly bent, under ether, and there seemed a strong probability that it would ultimately recover a great share of normal movement.

The case is an example of the difficulties that always seem to beset gunshot wounds, for in this category it may be ranked. In the first place the small size of the external wound deceived the naval surgeon, who first saw it, and who thought that the splinter of wood, which he removed, was all that it contained. In the next place the piece of metal, having buried itself in the outer condyle of the femur, set up a general ostitis of the lower end of the femur, and an inflammatory thickening of all the peri-articular tissues, together with the production of an abscess at the back of the joint. As the external sinus communicated with this abscess cavity, my finger went straight into it at the first examination, and it was not till many weeks afterwards that the second examination revealed the presence of bare and probably dead bone, in the course of searching for which the metal was discovered. It reminded one somewhat of Garibaldi's celebrated case.

Two Fatal Cases of Tetanus.—The first patient was a young healthy man of 22, whose right elbow got under the wheel of a cart, and was much lacerated and bruised by it. He attended as an out-patient at the Northern Hospital and Infirmary for eighteen days, and for some time seemed going on all right. But the wound began to look unwholesome, and the parts around the joint began to be swollen. So he was admitted as an in-patient, and on that very day exhibited the first symptoms of tetanus, by having a slight stiffness of the lower jaw. The parts were freely incised to give good drainage, but they never became wholesome, but always remained swollen and dusky, and discharged a thin, sanious fluid. The internal condyle of the humerus also became bare and carious. No good is to be got from trying a variety of plans of treatment, so a mixture of chloral and bromide of potassium was given from the first and persevered with during the whole progress of the case. The dose was five grains of chloral and ten of bromide of potassium, and this was given every two hours so that in twenty-four hours he got (with the two-hourly dose) 60 grains of chloral and 120 of the bromide. We had no doubt whatever that his tetanic spasms, which assumed the form of trismus and opisthotonos, were capable of being subdued by this treatment. More than once the medicine was given every four hours, and then the spasms became worse and were again subdued by a two hourly dose. They never were very bad, however, at any time, so that he was always able to swallow and breathe, and was not nearly so much distressed by them as by the pain of the bad arm. By careful feeding and this sedative treatment we kept him alive for sixteen days and at one time were very hopeful as to his recovery; but during the last week of his illness he got more and more feeble and eventually died in a semi-comatose condition.   No autopsy could be obtained.

The other case was much more severe.   The patient was a woman of 47, who sustained a severe laceration of the foot by its getting between a railway carriage and the platform.   On the morning of the seventh day after the injury trismus suddenly came on. Twenty grain doses of chloral were commenced and carried on all through, the intervals between each dose being shortened or extended according to circumstances.   On the afternoon of the seizure a tremendous attack of opisthotonos came on, accompanied by such respiratory difficulty that the patient became black in the face and almost died.   My colleague, Mr. Parker, being in the house performed tracheotomy, and put in a tube to prevent further risk of this kind.   The patient passed a quiet night with abundant sleep, so the chloral was much diminished; but on the evening of the second day the head became retracted and it was resumed.   On the third day there were no spasms, but the head was retracted, swallowing was almost impossible and the breathing very shallow, and on the fourth day a fatal termination took place.   The patient died sixty-seven hours after the first seizure, and during the first forty-eight hours two hundred and forty grains of chloral were administered.   The drug undoubtedly subdued the spasms and kept the patient comparatively free from pain; but the attack was a very severe one and the capacity for taking nourishment and for breathing were early interfered with, so that it cannot be said that any curative effect was perceptible.

After death a branch of the external planter nerve was found lying exposed among the sloughing tissues of the sole of the foot, but there were no naked-eye appearances of morbid change in it.

If we are to find any remedy for tetanus it must be done by selecting a drug and using it, and it alone, from the commencement, all through a series of cases.   So far as my limited experience goes, chloral or calabar bean seem to me to have given the best results—which is not saying much.   Chloral seems rather the better of the two, as regards the certainty with which spasms are subdued, so I mean to give it a fair trial.

VARIOUS CASES.

Nature.	Number.	Recovered.	Died.
Eczema, ...	4	3	1
Psoriasis,	1	1	0
Tinea tonsurans, ..	1	1	0
Coccydynia (operation), ...	1	1	0
Senile gangrene (foot and lung), ...	1	0	1
Charbon, ...	1	0	1
Cellulitis and necrosis of foot—chronic pyæmia, ...	1	0	1
Erysipelas,	1	1	0
Rheumatic pains—simulating joint disease,	2	2	0
Sinuses in old stump, ...	1	1	0
Hysterical affection of the shoulder muscles,	1	1	0
Empyæma—thoracentesis, ...	1	0	1
Ganglion of the wrist, ...	1	1	0
Syphilitic gumma of thigh,	1	1	0
Nasal polypi—injection of iron, ...	1	1	0
Pelvic inflammatory swelling (peri-uterine),	1	1	0

Coccydynia—operation.—The patient was a healthy young woman of 20, a domestic servant. She had suffered from pain in the coccyx for nearly two years, for which she had employed all sorts of local applications, without any effect. She thought the pain worst when she had a cold and it was certainly aggravated at the menstrual period. After walking it came on, and sitting was very uncomfortable—even coughing produced it. Pressure on the tip of the coccyx either from without or from the rectum, produced the pain at once. She was so wretched on account of the complaint that she was quite willing that anything should be done. Accordingly under ether, and with a finger in the rectum. a tenotomy knife was inserted at the tip of the coccyx and all the tissues attached to its sides completely divided. From that day the pain was relieved. The patient was seen a year after the operation and had remained well ever since. She was quite a sensible woman, who, when admitted into the Infirmary, was evidently in no little distress, and who agreed willingly to anything proposed for her relief, I therefore believe that there was undoubted pain; and undoubted and immediate relief from the operation. As to the pathology of the disease or the rationale of the *modus curandi*, I cannot say anything with certainty.

Mucous Polypi, treated by injection with perchloride of iron— Kate M'Grath, aged 21, had suffered from nasal mucous polypi for about three years, with all the usual symptoms of stoppage of the nostrils, loss of sense of smell, frontal headache and increased width of the bridge of the nose. I attacked one nostril with the forceps and brought away a small piece of turbinated bone and a group of little mucous polyps. But presently more made their appearance and the girl was so distressed by the prospect of another operation that I tried injecting the polyps with liq; ferri perchlor; fort; about three minims being sent into each polyp with a subcutaneous syringe. One would have imagined that mucous polyps were quite insensitive, but the injection always caused some pain. She had this done every second day for about eight or ten times, and the polyps shrank up so much under the treatment that the nostrils became free enough to enable her to use a nasal douche. She accordingly washed them out daily with a solution of iron and glycerine in water as strong as she could comfortably bear. She left the Infirmary able to breath through both nostrils, with the frontal headache quite gone and with the breadth across the root of the nose much diminished.

As to the most permanently curative operation for nasal mucous polypi I believe there is nothing equal to the use of the forceps properly managed. Where there are large, isolated polyps with well marked stalks the wire snare or Dr. Thudichum's process may do well enough—and probably removes them with much less pain than the forceps. But these are not the most common cases. On the contrary there are usually crops of small growths fringing the superior and middle turbinated bones, which no snare can get hold of, and which in due time make their appearance as large ones.

Mr. Syme, after great experience, used to say that the only way was to get one blade of the forceps beneath the turbinated bone and the other on the opposite side of it and carry away as much bone as possible. This I always endeavour to do, and find that along with the big ones I have brought away whole crops of minute polyps just commencing their existence, which can only be removed by carrying away the bone from which they grow. As to necrosis and all sorts of contingencies which it is said *may* occur as a result of such rough surgery, the simple answer is that they don't occur. On the other hand the patient has a chance of getting rid of the source of his trouble, and does not need to come every two or three years to have a fresh assault made upon a fresh lot. Failure often results from using forceps which are too big in the blades and which are only toothed at the points instead of all the way down. In not a few cases where the patient has had several operations performed previously by other surgeons, I have simply smashed up the whole turbinated bones as widely as I could, and so have settled the matter permanently. Now the pain and dreadful sensations produced by this proceeding are more than mortals can bear, and so the patients have had chloroform or ether; and it would be an excellent thing if this were resorted to more frequently. Even a moderate assault with the forceps is a most horrid process, and patients who have gone through it once or twice will endure any amount of chronic misery rather than face it again. But only a very few surgeons seem inclined to give these unfortunates an anæsthetic, urging as their reasons the danger of blood going down the throat and choking the patient and the fact that owing to the patient being insensible he cannot blow down the nostrils so as to let it be known whether they are clear or not. My plan is to have the patient thoroughly anæsthetized on a sofa. When fully insensible his head should be brought over the edge of the sofa so that the nostrils are dependant and then the surgeon, kneeling on the floor, passes up the forceps and pulls out everything he can till there is nothing more to pull. Meantime all the blood runs out of the nostrils and none need go down the throat at all, while the whole time necessary for a thorough clearing is about a minute for each nostril. I feel convinced that for certain cases the only satisfactory cure is to pull away as much as can be got of the superior and middle turbinated bones.

My friend Mr. Harrison some time ago sent to one of the medical journals a short notice of the plan of injecting with iron and so I tried it in the case just alluded to. It seems not a bad plan for nervous patients who object to any more severe operation, but obviously can only be of permanent benefit in cases where there are no crops of small polypi coming on.

Eczema impetiginodes—pneumonia—death.—Death during the progress of any of the more ordinary skin diseases is very rare, however severe the attack; which may make the present case worth noting. The patient was a poor, miserable dressmaker, aged 17, who had been very badly fed for some time before her admission. The eruption was all over her face and head, and upon her hands. On the face and head it was very severe and looked as if it were going to be complicated by erysipelas—the eyes being closed up and the neck and ears terribly red and swollen. Under treatment this began to die away, and about ten days after her admission the eruption appeared on the belly in an isolated pustular-looking form. Her temperature then began to rise and she shewed all the signs of pyrexia for which nothing was found accountable, until a careful examination of the chest revealed dulness at the base of the right lung. During the next five days she had slight cough, and a temperature running between 102 and 105, with delirium and great restlessness, which ended in diarrhœa, insensibility and death. Digitalis was tried in small hourly and two-hourly doses in the hope of reducing the pulse and temperature, but it was not found of any real service. The fact was that the girl had been so badly nourished that she had no stamina to stand up against a severe febrile condition. After death there was found a little recent pleurisy over the base of the right lung, which was the subject of acute lobar pneumonia, under which it was almost in a breaking-down condition. From its cut surface pus could be scraped in considerable quantity. The other organs were all healthy.

SOME NOTES ON ETHER AND CHLOROFORM.

TWELVE years ago I should think there was hardly a score of middle-aged surgeons in England who had seen ether administered, although of course the seniors remembered it in the earliest days of anæsthetics. To-day it has elbowed chloroform out of the field, and the journals have teemed with aggressive letters from rabid etherists and defensive effusions from prejudiced chloroformists. Liverpool being practically the landing-place for Americans, we are constantly favoured at the Infirmary with visits from American surgeons, and were very early initiated into ether-giving. In 1872, when on a visit to the States, I first saw it administered at the Massachussett's General Hospital, and in the following year an American surgeon, whose name I now forget, etherized some patients for us at the Infirmary and gave us our first practical instruction. He saturated a big cone sponge with the anæsthetic, and then, squeezing the superfluity on the floor, clapped it over the patient's nose and mouth, and, putting a towel on top of it, held it on in spite of the patient's splutters and struggles till he was quiet. Although this somewhat garroter-like method of administration was speedily given up, ether at once found favour and has been the general anæsthetic in use here ever since. Our experience of it is therefore probably as long and as extensive as that of any other institution in the country.

From the teaching of Simpson I naturally imbibed an unquestioning belief in chloroform, while for six months, as a senior student, I administered it in the theatre of the Edinburgh Royal Infirmary to all Mr. Syme's patients. From the latter I received only two instructions—(1) Give plenty of chloroform and plenty of air; (2) watch the breathing most jealously, and pull out the tongue at once if it is not absolutely free and full. To this day I believe that these are the golden rules. If a patient is going to die from the heart, pure and simple, the failure is so sudden and the catastrophe so complete that no man can avert it. *The pulse in the great majority of cases does not give the warning that might be expected from it.* Therefore it is that with chloroform a certain number of deaths are absolutely unavoidable; but out of all the deaths that occur these, I firmly believe, form only a small minority. It may be a harsh thing to say (I have said it, however, before and will not retract), but I cannot help thinking that most of the fatal cases result from an inability on the part of the chloroformist to appreciate the exact condition of his patient and recognize impending danger, and I adhere to the doctrine which I was first taught that, in the fatalities which might have been avoided, a neglect of the breathing has been the main error. Of all the scores of medical students whom I have helped to educate only a limited number seem ever to grasp this question of breathing. Times without number I have seen a house surgeon solemnly pouring on the chloroform long after the patient's tongue had fallen back, and while for a minute or more he had not been getting into his lungs as much air as would make two good breaths. Some one has stepped forward and lugged out the tongue with a pair of forceps. Then has come such an inspiration as shewed what the unfortunate man was literally dying to get could he only have spoken. During an operation of a quarter of an hour a patient is often ten minutes on quarter rations of air—the working of his chest and belly deceiving the inexperienced administrator into the belief that his breathing is going on quite well. If you held a man's head under water till he was half drowned and then proceeded at once to give him a considerable dose of chloroform, would you be surprised if unpleasant symptoms appeared? And yet it is considered wonderful that they should appear after he has been half choked on a table! It may be asserted that there ought to be no difficulty whatever in recognizing the state of a patient's breathing:—but there is. If the patient is not making a noise with his throat it is sometimes far from easy to know how much air is going in or out. I always put my hand in front of the patient's mouth and nose and feel his breath on it—simple, but very certain.

All sorts of reasons are given to explain why the breathing becomes impeded. I stick to the old-fashioned notion that it is simply because the tongue falls back. Why does a drunken man, when he is put on his back, snore and grunt, and gasp till you are glad to roll him on his side? Simply because he has lost control over his tongue. And what is a chloroformed patient but a very, very drunk man? It has been argued that the pulling up of the tongue produces a reflex act of breathing. I see no necessity for any such explanation. It simply opens up the way for the air to get down. In an operation for removal of the tongue, mentioned at page 65, I performed a preliminary laryngotomy. The patient, having come to breathe quietly through his tube, the tongue was hauled out and dragged about to get ready for the cutting. I noted particularly that the quiet breathing through the tube was not altered in the slightest degree. If pulling out the tongue excites convulsive reflex acts of breathing why did this man not gasp through his tube? For the very good reason, I suspect, that this reflex view is a theory. Then again it has been asserted that dissections prove that the tongue cannot fall back so as to block the air way. Some years ago I was a member of a committee appointed by the Liverpool Medical Society to inquire into the merits of the various systems of producing artificial respiration. I was deputed to make dissections to ascertain about the position of the tongue; but after one or two attempts I found that it was utterly useless to compare the stiff and rigid muscular condition of a cadaver with the conditions during life. Even after the first rigidity has gone off, dead muscles cannot be compared with living ones, even although these living ones belong to an insensible man. I gave up the attempt as being quite incapable of giving results to be depended upon, and put no faith in the reports of dissections upon this matter. There is but a solitary objection to pulling a patient's tongue out, and that is that it makes it a little sore. On the strength of this all sorts of protests have been entered against what has been termed this "cruel and barbarous practice." This is twaddle to begin with; but in truth the tongue need not be hurt at all. If a pair of ring forceps be used the patient need never know that his tongue has been meddled with.

I am glad to say that I have only seen one death from chloroform, but as it occurred during the period to which the present cases refer, some particulars of it may be worth giving. The patient was a big framed, sallow man of about 35, a musician, and not an intemperate man so far as we could learn. He suffered from disease of the ankle-joint. To explore the articulation ether was given, but he was so obstreperous under it and got so choked with frothy mucus that chloroform was substituted in order to conduct the examination quietly. After a while Syme's amputation was performed. He had chloroform given at once this time and

took it well.   The stump never properly healed and after the lapse
of many months it was clear that the lower end of the tibia was
carious.   He therefore re-entered the Infirmary and I opened up
the stump and sawed off a good slice of tibia which was diseased
and full of cheesy stuff.   Chloroform was again given on the
ground that he had already taken it twice without any trouble,
while with ether he had been very unmanageable.   The operation
lasted about twenty minutes.   He had a full amount of chloroform
given on a flannel mask.   During the operation, although insensible
to pain, he struggled a good deal at intervals and shouted.   The
operation was just concluded, when, the tourniquet being removed,
I noticed that no blood spirted from the vessels and was surprised
thereat.   Almost immediately it was clear that something had gone
wrong with the anæsthetic.   Instantly going to his head I pulled
out his tongue, but, to my dismay, his jaw was flaccid and dropped,
his eyes were staring open, his pupils were widely dilated and not
a trace of pulse was to be felt.   He was dead.   Not realizing this,
however, we inverted him to let the blood run into his brain—
then I rapidly opened the trachea and we employed Sylvester's
method of artificial respiration, with galvanism and all the ordinary
remedies.   It was useless.   From the moment that the house
surgeon first saw that something had gone wrong, the patient never
shewed the slightest sign of life.   When we came quietly to think
over the incidents of the case it was remembered that the man was
talking and moving involuntarily not more than two to three minutes before the alarm was given.   My house surgeon, a very careful man with anæsthetics, observed nothing wrong with the breathing, but had his attention first drawn to the patient's dangerous
state by his extreme pallor and by the fact that his jaw dropped.
After death it was found that he had a partially adherent pericardium; but not much else amiss either with the heart or other
organs.   However, fatty heart and adherent pericardium are
certainly the two most dangerous lesions as regards chloroform, and
they are the very things that the usual cursory stethescopic examination fails to detect.   As to ordinary valvular disease I do not
consider it any bar whatever to the administration of anæsthetics
if carefully watched.   As regards the fatal case just mentioned it
seemed to be a genuine case of death from paralysis of the heart,
as proved by the almost instantaneous nature of its occurrence, and

the perfect failure of all attempts to elicit even the faintest symptom of vitality.

My view then is that most of the fatal cases result from a neglect of the breathing, by which the patient's blood becomes so carbonized from want of air that the chloroform poison arrests the action of the respiratory centres with ease. The unavoidable cases are those where the poison acts suddenly upon the cardiac centres, which succumb at once and for ever. As it is impossible to pronounce with anything like certainty, whether an individual's heart is likely to give way or not, we have here an argument—an irresistible argument—for the general use of ether. The respiratory difficulty is to be prevented by keeping the tongue well pulled out; —is to be encountered when it does occur, by immediately opening the larynx or trachea, and artificially inflating the lungs by Sylvester's method and by blowing into a tube thrust into the windpipe. The cardiac difficulty is to be dealt with by inverting the patient and letting the blood run into his anaemic brain. Has enough been done, by the way, with nitrite of amyl?

As according almost entirely with the results of one's clinical experience, I was much pleased to note some time ago, in the *Lancet*, a synopsis of the experiments of M. Vulpian, on animals, laid before the Paris Academy of Medicine.—"The medulla oblong-" ata has been thought to be unaffected by the inhalation of chloro-" form, but, although resisting the effects of chloroform longer than " other parts of the nervous system, it is unquestionably affected in " some degree, even in moderate anæsthesia. If the pneumogastric " nerves are divided in a healthy animal, it continues to breathe. " Stimulation of the central ends then arrests respiration, which " recommences, even although the peripheral extremities are fara-"dized. But, if the same experiment is performed on an animal "under the influence of chloral or of chloroform, the breathing does " not recommence after it has been arrested by stimulation of the " divided pneumogastrics. This shews that in the anæsthetic state "the medulla oblongata is in a functional condition different from "its normal state. Again, in an animal under normal conditions ' faradization of the peripheral ends of the divided pneumogastrics "arrests the heart in diastole; but, if the stimulation is continued " the contractions recommence. If the same experiment is per-" formed on an animal under the influence of chloroform, the arrest

" of the heart is more readily produced and is final.   Hence chloro-
" form acts on the respiratory centres; but, it acts also upon the
" motor ganglia of the heart.   In animals the cardiac failure is
" much more grave than the respiratory failure; life cannot be
" saved in more than one in forty of the former, although in the
" more frequent cases of respiratory failure life can often be pre-
" served by artificial respiration.   M. Vulpian fully confirms the
" slighter degree of danger involved in the use of ether, in conse-
" quence of which he invariably prefers it as a means of obtaining
" anæsthesia in experiments on animals."

Our early experiences here soon convinced us of the stimulating
effect of ether on the circulation, compared with the depressing effect
of chloroform.   This of course has been rigidly demonstrated beyond
a doubt with the aid of sphymographs, cardiographs, haemadynamo-
meters and all sorts of scientific machines.   Quite right: but any-
one who has seen the ruddy face and rosy lips of the etherized
patient and compared them with the pallid hue of the chloroformed
one needs no further experiment.   Is there then no danger with the
etherized man? Undoubtedly there is; and it would be a very great
thing if house-surgeons and others would bear in mind that no
patient can be rendered insensible by any drug whatever without
some danger.   One of the recorded deaths from ether occurred at
one of the hospitals in this city; but it was due to the fact that the
patient having recently had his dinner, was sick during the operation, and giving a violent inspiration sucked down into his trachea
and bronchi such an amount of vomited matter as fairly choked
him.   But this is a preventable occurrence and cannot be ascribed
to the anæsthetic.   The chief risk is undoubtedly with bronchitic
patients in whom ether produces such a rapid and excessive secretion of mucus that the air tubes are completely filled with it.   I
remember one instance in private practice where the patient during
the operation seemed in very considerable danger, and gave great
anxiety by her lividity and embarrassed breathing to a most experienced etherist.   Moreover not only is the difficulty at the time of
operation, but a patient with a cough has it decidedly aggravated
by ether, so that in a case, where it is particularly desirable that
there should be no coughing or straining after operation in a
bronchitic patient, ether is certainly not good.   As regards sickness
I cannot say I have seen much difference between it and chloroform

during the administration. Certainly it cannot claim any advantage—but after operation there is a marked advantage. I have never seen with ether the horrible and prolonged sickness, lasting over one or two days, which was not so very uncommon with chloroform.

As regards minor difficulties the only one of moment which we have noted here is that with ether, it is sometimes very difficult thoroughly to control muscular movement in powerful, full-blooded men. This continues long after ordinary sensibility to pain is abolished. So that one might cut a patient's arm off without his knowing it, while it would hardly be safe, owing to his wriggling and struggling, to do a delicate operation such as the ligature of a large vessel. The anus and urethra also seem to be a long time in being quite deadened with ether. I have more than once been quite beaten at an operation for hæmorrhoids, owing to the patient drawing up his rectum and pushing down his legs as soon as the orifice was touched. I remember one day my colleague Mr Harrison was performing Bigelow's crushing operation for stone. Ether was poured on *ad lib;* and time abundant for action given several times, but no sooner was the lithotrite in the bladder than such a straining and wriggling took place as rendered careful search for fragments impossible. Chloroform very soon settled all this. So that there is little doubt that while ordinary sensibility to pain disappears soon enough under ether, that drug takes longer than chloroform to paralyze reflex action. So much the better in one sense—the more difficult is it to paralyze the heart with it.

As regards inhalers for anæsthetics they cannot be too simple. In Liverpool a handkerchief, a towel or a piece of flannel shielded over a wire frame are the ordinary vehicles—and none better. For ether I prefer the American wire apparatus, with the stocking bandage folded through it and surrounded with india-rubber. A great many inhalers have recently come out whose purpose is to save ether and rapidly render the patient insensible. They are all made on one common principle, that of making the patient breathe into a bag and re-breathe the same air over and over again. They all achieve their two objects of saving ether and rapidly anæsthetizing the patient, by putting the patient in the most dangerous position possible for any accident to happen. They do it by simply depriving him of oxygen ; by carbonizing his blood.

Exactly the same result can be got by compressing the internal jugular veins.   If any accidents happen with ether, I feel sure it will
be with some of these machines.   In order to save a shillingsworth
of ether a patient will be so admirably carbonized some day that
the drug will overpower his respiratory or cardiac centres.   Then
the anæsthetizer will write a long letter to the medical papers giving the amount of ether used and a host of other details, together
with an admirable report of the autopsy ; but not mentioning that
during the whole of the administration he had been doing his best to
diminish the patient's powers of resistance by quietly suffocating
him.   Plenty of anæsthetic and plenty of air sums up my
creed in this matter.   As for those elaborate apparatuses, by
which regulated quantities of air and anæsthetic are turned on, they
are doubtless admirable, but a country practitioner in the wilds of
Cumberland cannot travel about with an affair on his back like a
Parisian gingerade seller.   They may be all very well for fashion-
able ladies in the houses of eminent London dentists, but the *hoi
polloi* have to get their ether and chloroform too, and the best
apparatus for them is the simplest.   Fortunately in surgery the
best weapons *are* always the simplest.   It is the workman, not the
tool.

I am almost ashamed to find how far my lucubrations have led
me, and begin to think that I have been making much ado about
nothing.   However, I have a rooted dislike to the orthodox and
strictly proper method of reporting cases.   First comes the pre-
amble, which, by reason of being put in one fixed and unaltering
form, makes every case as like another as two peas.   Then generally follow the entrees, prepared as follows:—"Jan. 4th, patient
passed a bad night.   Temp; this morning 100°.4, pulse 98, resp;
21; bowels moved three times; partook freely of nourishment."
Followed by a statistical *pièce de resistance*, only to be digested by
those gentlemen who have to write dictionaries and text books;—
for whom it is very nourishing.   Very little dessert.   I hope that
the friends who may skim over these pages will pardon me for
having served up my cases after my own fashion, and for having
put before them nothing more than a small dish of surgical gossip;
—which is all this pamphlet professes to be.—**W. M. B.**

www.ingramcontent.com/pod-product-compliance
Lightning Source LLC
Chambersburg PA
CBHW021707210326
41599CB00013B/1554